从概念到建筑
From Concept to Building 1

佳图文化 编

华南理工大学出版社
SOUTH CHINA UNIVERSITY OF TECHNOLOGY PRESS
·广州·

图书在版编目（CIP）数据

从概念到建筑．1：英汉对照 / 佳图文化编．— 广州：华南理工大学出版社，2014.8
ISBN 978-7-5623-4269-4

Ⅰ．①从… Ⅱ．①佳… Ⅲ．①建筑设计—作品集—世界—现代—英、汉 Ⅳ．① TU206

中国版本图书馆 CIP 数据核字（2014）第 110054 号

从概念到建筑1
佳图文化 编

出 版 人：	韩中伟
出版发行：	华南理工大学出版社
	（广州五山华南理工大学17号楼，邮编510640）
	http://www.scutpress.com.cn E-mail: scutc13@scut.edu.cn
	营销部电话：020-87113487 87111048（传真）
策划编辑：	赖淑华
责任编辑：	赵宇星 陈 昊 孟宪忠
印 刷 者：	广州市中天彩色印刷有限公司
开 本：	787mm×1194mm 1/16 印张：20
成品尺寸：	230mm×300mm
版 次：	2014年8月第1版 2014年8月第1次印刷
定 价：	398.00元

版权所有　盗版必究　　印装差错　负责调换

PREFACE 前言

From concept to building is not a process of simple duplication or deliberately piling up, but a transformation of abstract beauty in concept to beauty of building in form, an expression of conceptual connotation to concrete beauty of building. From concept to building is a process to realize our vision of a better city in line with the law of beauty.

According to the latest design concepts in current international construction industry and from the perspective of professional architectural design, this book is a collection of carefully selected cases mainly involved in shopping centers, commercial complexes, urban complexes, office buildings, hotel buildings, art and cultural buildings and transportation buildings and so on. In content layout, each case is analyzed in keywords, features and design concept in the company of a large variety of technical drawings such as plans, sections and analysis graphics. With rich content and full and accurate information, the book tries to give architects and readers in related industries visual enjoyment and design inspiration.

从概念到建筑，不是简单复制的过程，也不是刻意堆砌的过程，而是将概念的抽象美转化成建筑的形式美、将概念的抽象内涵转化成建筑的具象表达。从概念到建筑，是按照美的规律，实现"我们让城市更美好"的愿景的过程。

本套书依据国际现行建筑行业的最新设计概念，站在建筑设计的专业角度，精心挑选案例。本书案例主要涉及了购物中心、商业综合体、城市综合体、办公建筑、酒店建筑、文化艺术建筑以及交通建筑等建筑形态。内容编排上，分别从案例的关键点、亮点、设计概念等方面入手，并配合大量的各种技术图纸，如平面图、剖面图、分析图等。本书内容丰富、资料详实，希望能给建筑设计师及相关行业读者带来视觉享受和设计启迪。

CONTENTS 目录

Shopping Center 购物中心

002	Jining Jiulong Guihe Shopping Plaza	济宁九龙贵和购物中心
008	Qingdao Wonderful · Guan Jing Shang Pin	青岛万德丰·观景尚品
014	Hefei SASSEUR Art Commercial Plaza	合肥砂之船艺术商业广场
028	Yibin NHH Fortune Center	宜宾唐人财富中心
038	Shenyang Chanceave City Mall	沈阳千姿汇购物中心
042	Star City International Plaza	城际星港城
046	Penglai Joy Plaza	蓬莱悦动港湾购物街区
054	Commercial Plaza on Nanjing Road, Tianjin	天津南京路商业广场
066	Fengdong International	大明宫·沣东国际
078	Shanghai Fengxian Nanqiao Powerlong City Square	上海奉贤南桥宝龙城市广场
090	Kunming Nanshi Central Golden Estate Project – Phase II	昆明南市中央金座二期
096	Qingdao Zhengjian Golden Town	青岛政建金地世纪城
104	Shanghai Jiading New City Plot A131	上海嘉定新城A131地块
118	Nanhai Xinji Plaza	南海信基广场
130	Changsha Shimao Plaza	长沙世茂广场
134	Jinan Dinghao Plaza	济南丁豪广场
146	Business Street in Huishan District, Wuxi	无锡惠山区商业街
154	Yingtian Road Market Center in Nanjing	南京应天大街营房仓库商业开发

Urban Complex 城市综合体

164　Core Casa, Tianjin　　天津滨海新城

168　Taihe Urban Complex, Fuzhou　　福州泰禾城市综合体

176　Mojiazhuang Commercial Plot, Wuxi　　无锡莫家庄商业地块

186　Trans-surface, Beijing　　北京"超表皮"

190　Delta Area of Jintang County, Sichuan　　四川金堂县三角洲、三星片区

196　Minmetals International, Tianjin　　天津旷世国际

204　Wanlong Shenyang (Shenbei) Project　　沈阳万隆（沈北）项目概念设计

216　Poly Real Estate's Yudong District Project Design　　保力地产御东区项目规划设计

222　Reconstruction of Luan County Guangming Plaza　　滦县光明商城改造

230　Quanshun Fortune Center Plot 14　　泉舜财富中心14#地块

236　Lanyue Bay West Convention Centre & Mixed Development　　揽月湾西地会议中心及综合发展

246　Guangzhou International Bioisland　　广州国际生物岛

252　Liuzhou Customs Port　　柳州风情港

260　CITIC Jinluan Bay, Zhangzhou　　漳州中信东山岛金銮湾

266　Fantasia·Long Nian International Center　　花样年·龙年国际中心

272　Shijiazhuang Urban Complex　　石家庄中委城市综合体

278　Chengdu Lingdi International Plaza　　成都领地国际广场

282　Administrative Center in Fengdong New Town, Xixian New Area　　西咸新区沣东新城管委会政务中心

286　Civil Aviation Air Traffic Control Base Logistics Service Center　　民航空管基地后勤服务中心

290　Langxi Guogou City　　郎溪国购城

302　Summer International Retail and Entertainment Center　　世邦国际商贸中心

308　Qingdao Urban Balcony　　青岛城市阳台

Shopping Center
购物中心

High Quality
突出品质

Artistic Atmosphere
营造艺术

Facade Design
立面形象

Space Organization
空间组织

Jining Jiulong Guihe Shopping Plaza
济宁九龙贵和购物中心

Keywords 关键词

Elegant Shape
造型简洁

Interwoven Blocks
体量穿插

Green System
绿化系统

Location: Jining, Shandong, China
Developer: Jining Jiulong Guihe Shopping Plaza Co., Ltd.
Land Area: 31,987 m²
Floor Area: 149,189 m²

项目地点：中国山东省济宁市
开 发 商：济宁九龙贵和购物广场有限公司
占地面积：31 987 m²
建筑面积：149 189 m²

Features 项目亮点

The building looks modern and elegant with huge blocks interweaving with each other to create various interesting spaces. The terraces on the top of the podium become roof gardens on different levels, forming a three-dimensional green system.

整个建筑造型简洁时尚，大体量相互穿插，自然形成了多种趣味空间。裙房顶部的层层退台则成为不同标高的屋顶花园，形成立体的绿化系统。

■ Overview

Situated in the center of Jining commercial area, the project enjoys great development potential. Its land area is 31,987 m², and the floor area is 149,189 m². It will be a typital, large center for commerce, entertainment, and relaxation in the area.

■ 项目概况

该项目居于济宁市商业区中心位置，具有良好的市场发展前景。整个项目的用地面积为31 987 m²，建筑面积为149 189 m²。建成后将成为区域内具有代表性的集大型商业、娱乐、休闲等为一体的综合中心。

■ Planning

Based on the long history and natural resources of Jining City, making full use of the tourist advantages, and exploring the favorable conditions of the site, it will created a new landmark complex for commerce, relaxation, experience and hotel with high starting point, high standard and human care. Meantime, it will be a platform for relaxation, entertainment and shopping. This new commercial model will bring the local residents with great shopping and entertainment experience.

■ 项目规划

项目规划依托济宁的悠久历史和良好的自然资源，发挥休闲旅游的先天优势，充分发掘本地块的良好资源，打造高起点、高标准、以人为本的集商业、休闲、体验、酒店居住为主的综合性新地标，同时，为市民提供一个更加适合休闲、娱乐、购物的生活交际平台，打造一种当地全新商业模式，带给当地居民前所未有的购物休闲娱乐之旅。

■ Architectural Design

The whole building looks simple and fashionable. Big blocks interweave with each other to create varied interesting spaces. The terraces on the top of the podium have formed roof gardens of different heights. Thus a three-dimensional green system is formed. In terms of details design, the project is 45m away from Jinyu Road. People are led into the underground floors by sunken plaza to realize the layout of "double first ground". It has increased the commercial value and enriched the frontage space to create a unique entrance space. On the east side, auxiliary road is set to release the traffic pressure and ensure the smooth of the main urban road. Different functional blocks are independent without mutual disturbance. Meantime, it has provided enough outdoor spaces for the hotel.

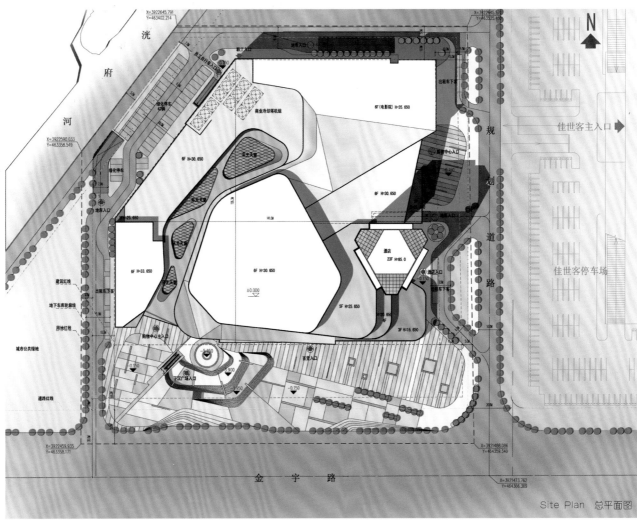

Site Plan 总平面图

003

■ 建筑设计

整个建筑造型简洁时尚，大体量相互穿插，自然形成了多种趣味空间。裙房顶部的层层退台则成为不同标高的屋顶花园，形成立体的绿化系统。在细节设计上面，项目沿金宇路退让45 m，利用下沉式广场将人流引入地下，带旺了地下楼层，实现"双首层"的格局，提高商业附加值，同时丰富了沿街步行空间的形式，营造出别具一格的入口空间。东侧在地块内设置辅道，缓解了商业车流对城市道路造成的交通压力，维持了主干道车行顺畅。各功能块流线独立成区，互不干扰，同时也为酒店提供了充足的室外活动空间。

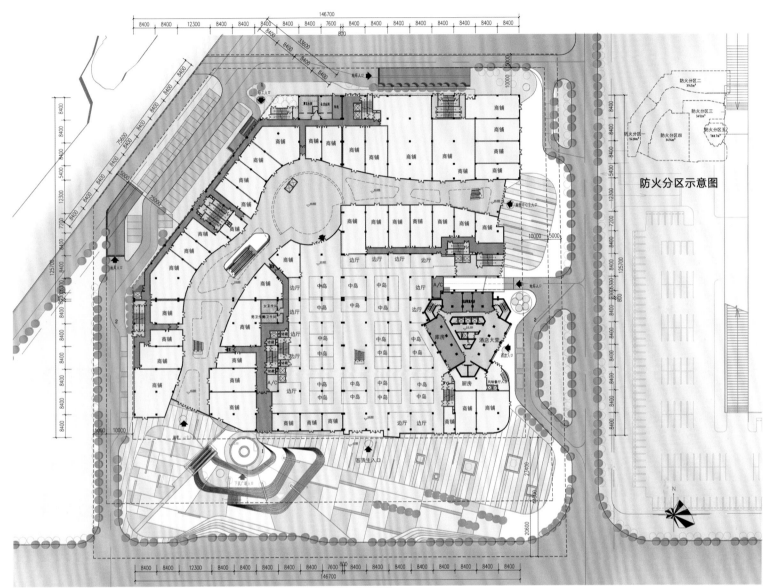

First Floor Plan 一层平面图

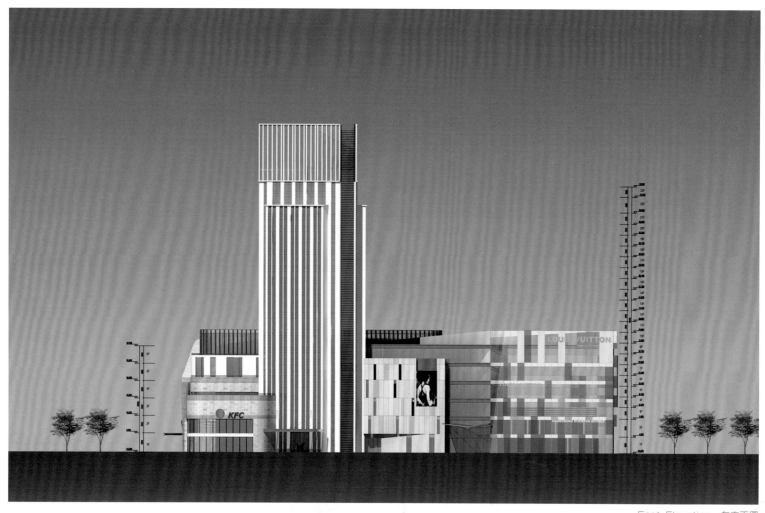

East Elevation 东立面图

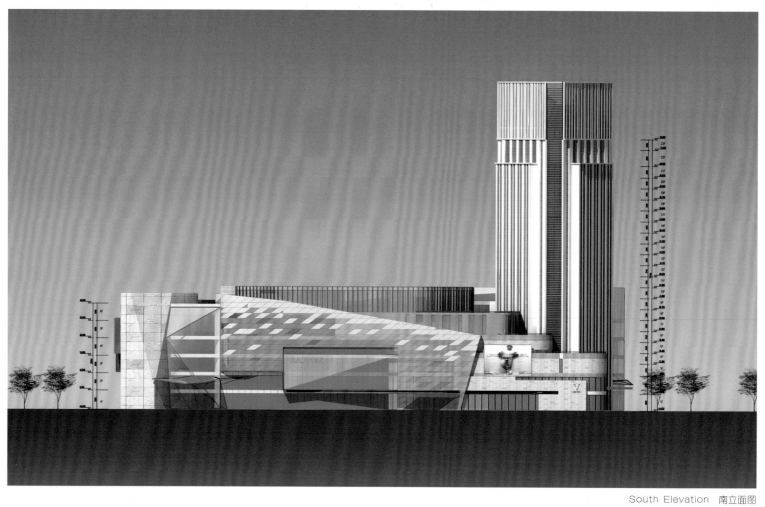

South Elevation 南立面图

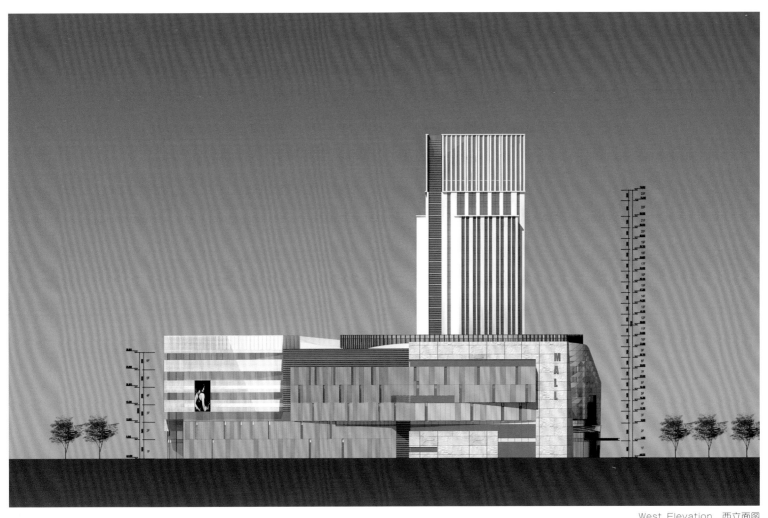

West Elevation 西立面图

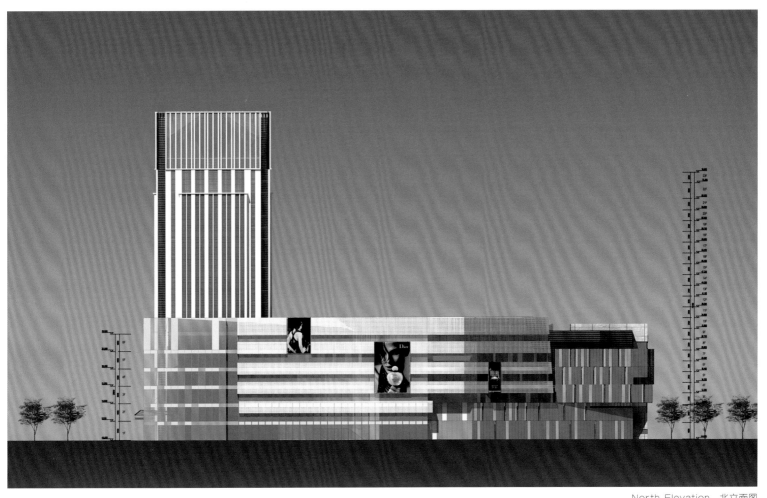

North Elevation 北立面图

Qingdao Wonderful · Guan Jing Shang Pin
青岛万德丰·观景尚品

Keywords 关键词

Glass Facade
玻璃立面

Elegant Appearance
造型优美

Reasonable Arrangement
错落有致

Location: Qingdao, Shandong, China
Client: Qingdao Wonderful Industrial Development Co.,Ltd.
Architectural Design: Tengyuan Design Institute Co.,Ltd.
Site Area: 8,573 m²
Total Floor Area: 26,702.8 m²
Overground Floor Area: 19,718 m²
Underground Floor Area: 6,984.8 m²
Plot Ratio: 2.3
Green Coverage Ratio: 30.01%

项目地点：中国山东省青岛市
业　　主：青岛万德丰实业发展有限公司
设计单位：青岛远建筑设计事务所有限公司
用地面积：8 573 m²
总建筑面积：26 702.8 m²
地上建筑面积：19 718 m²
地下建筑面积：6 984.8 m²
容 积 率：2.3
绿 地 率：30.01%

Features 项目亮点

It pays attention to create specific atmosphere for this development. Buildings are arranged with proper distance and connecting with each other in order.

本项目设计重点是对场所的营造，充分考虑建筑之间的空间，使得这些公共空间互相连接，形成体系。

■ Overview

Situated on the east of Tangdao Bay of Qingdao Development Zone, within the Shiquetan Residential Area, the development is ideally located with beautiful sea views on the west and golden beaches to the east.

■ 项目概况

本项目位于青岛市开发区唐岛湾东侧，石雀滩居住区内，西向面临唐岛湾海景，东侧与金沙滩景区遥望，自然景观条件极其优越。

■ Design Goals

1. Maximizing the interface along the street to upgrade the commercial value.

2. Preparing for the management and operation of the commercial properties in the future.

3. Paying attention to create specific atmosphere with the buildings connecting with each other in proper distance.

4. Creating vigorous public spaces for living and relaxation and giving people unforgettable space experience.

■ 设计目标

1. 最大化地创造沿街界面，从而提升商业产品的价值。

2. 做好后期商业物业运营的准备，以迎接商业运营带来的威胁。

3. 本项目设计重点是对场所的营造，充分考虑建筑之间的空间，使得这些公共空间互相连接，形成体系。

4. 它将包含富有生机的公共空间，为人们提供游憩休闲空间，它将提供给人们独特的居住生活以及游览体验，使人流连忘返。

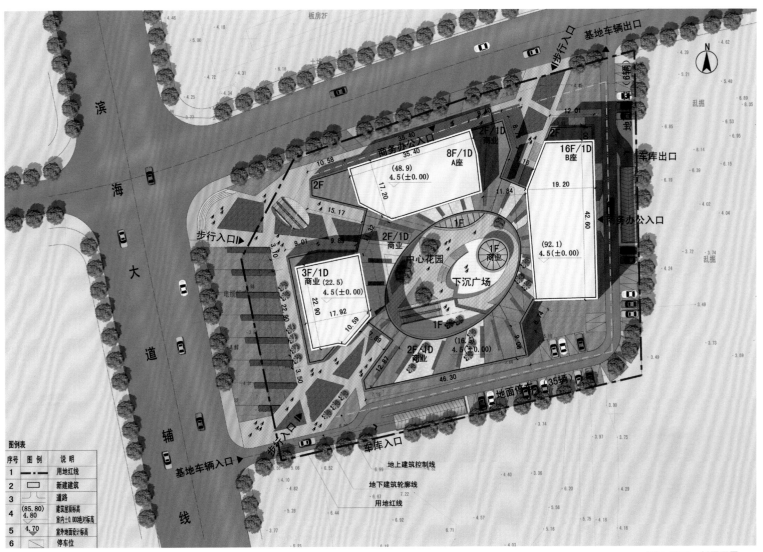

Site Plan 总平面图

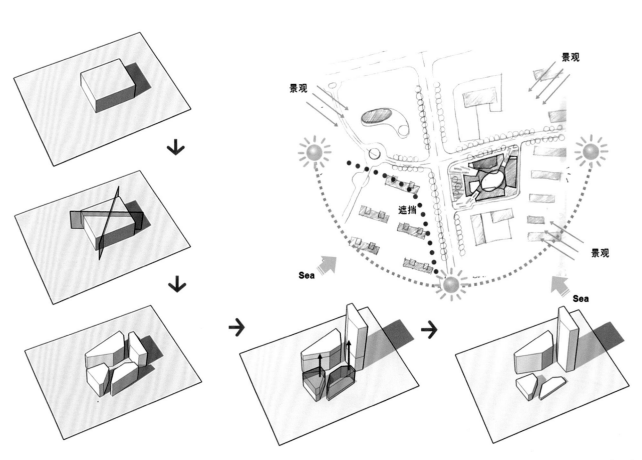

Scheme concept 方案概念

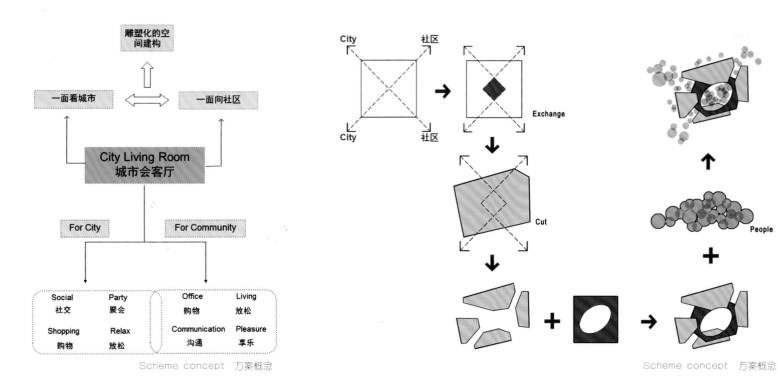

Scheme concept 方案概念

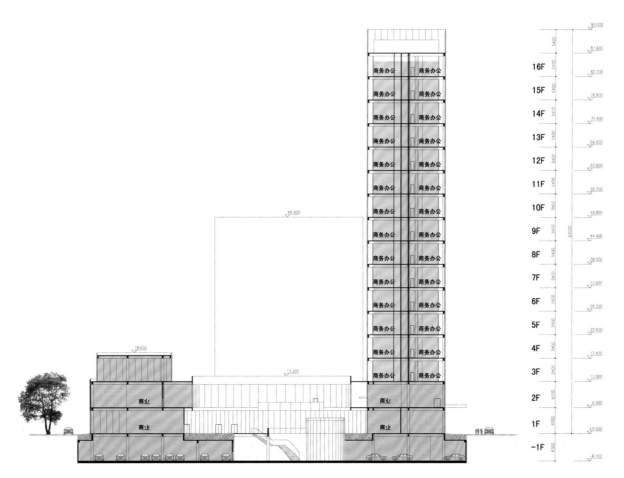

Section 1-1　1-1 剖面图

Hefei SASSEUR Art Commercial Plaza
合肥砂之船艺术商业广场

Keywords 关键词

- European Style Architecture
  欧式建筑
- Tile Collage
  面砖拼贴
- Super-Scale Arcade
  超尺度拱廊

Location: Hefei, Anhui, China
Architectural Design: China Construction International (CCI) Architecture Design & Consulting Co., Ltd.
Planning Site Area: 132,212 m²
Total Floor Area: 316,106 m²
Plot Ratio: 1.70
Green Coverage Ratio: 20%

项目地点：中国安徽省合肥市
设计单位：上海新外建工程设计与顾问有限公司
规划用地面积：132 212 m²
总建筑面积：316 106 m²
容 积 率：1.70
绿 化 率：20%

Features 项目亮点

For the building facade, the main building Outlets adopts the Swiss architect Mario Botta's architectural style and it makes wonderful extraction on the traditional European architectural language and applies facade processing technique of tile collage, which looks concise with rich details, and through the comparing of the scale and materials, it highlights the texture sense and taste of the building.

在建筑立面上，主体奥特莱斯采用瑞士建筑大师马里奥·博塔的建筑风格，对传统的欧式建筑语言进行精彩的提炼，采用面砖拼贴的立面处理手法，简洁中富含丰富的细节，通过尺度、材料的对比，突显建筑质感与品位。

■ **Overview**

The project is located in the east of Changning Road of National High-Tech Industrial Development Zone in Hefei Shushan District, the south of Rainbow Road, and the north of Wangjiang Road. The project base presents like a rectangle about 500 m long in the south-north direction and 350 m wide in the east-west direction; it has an elevation above sea level between 38 m to 47 m, high in west and low in the east with moderate elevation difference on the whole.

■ **项目概况**

项目位于合肥市蜀山区国家高新技术产业开发区长宁大道以东，彩虹路以南，望江西路以北。基地基本呈南北长约500 m、东西宽约350 m的长方形，海拔在38~47 m之间，整体呈西高东低，地形高差适中。

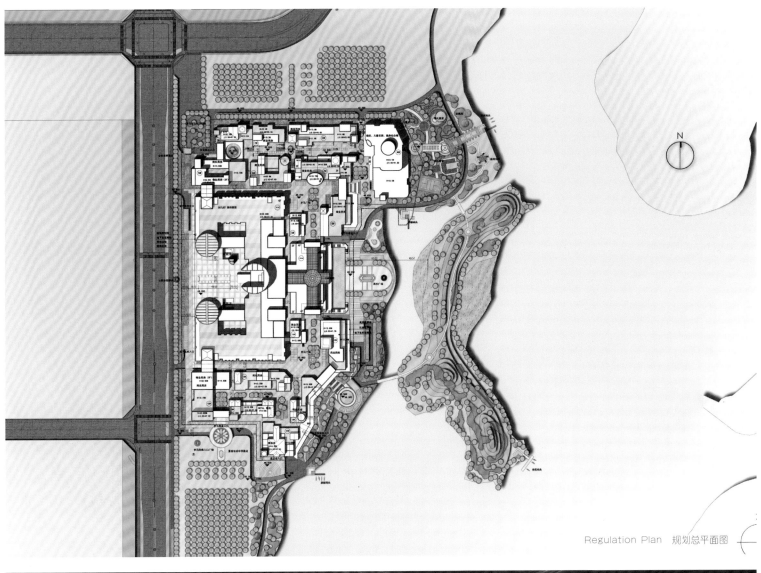

Regulation Plan 规划总平面图

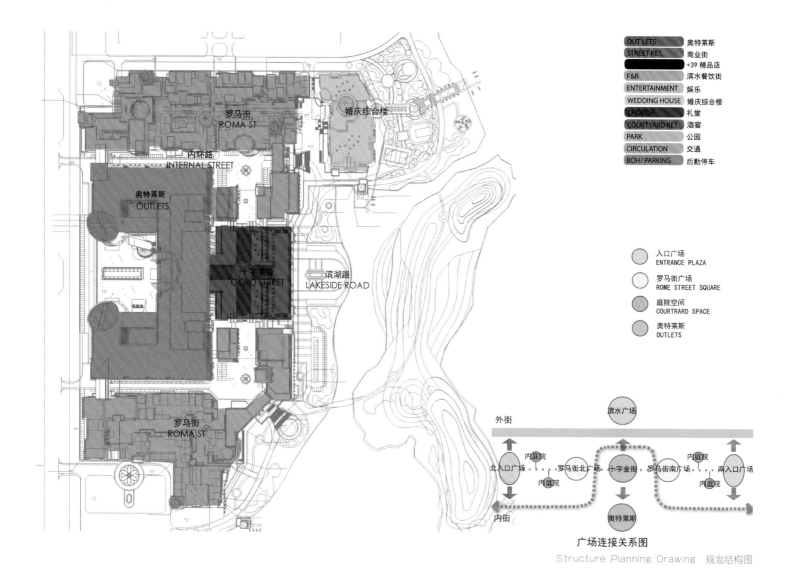

Structure Planning Drawing 规划结构图

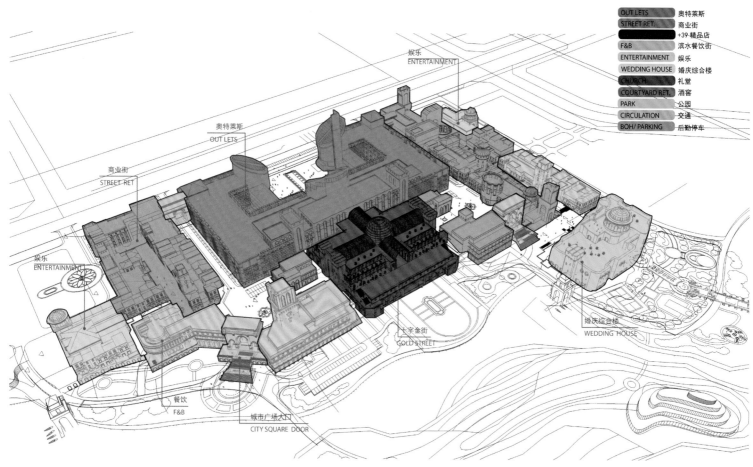

Functional Analysis Drawing 功能分析图

■ Planning Concept

The project planning concept is "one axis, one center, double loops and six nodes":

"One axis" is constituted by a main plaza along the Changning Road, Outlets Theme Building, Cross Gold Street, Waterfront Plaza and Wangju Lake in the east-west direction, which is not only the main business line, but also a landscape axis.

"One center" refers to the core building group made up of the main Outlets building and Cross Gold Street; as the key space of the whole project, it brings together all kinds of international brands, which becomes the central node and spiritual space of the entire project.

One of the "Double loops" is formed by the main Outlets building and the Roman Customs Street, which is on the main internal motor vehicle flow. Another loop is formed by the Roman Customs Street itself, serving as the main shopping pedestrian flow.

"Six nodes" refers to six landscape spatial nodes, including the plaza in front of the main Outlets Building near Changning Road, south and north Roman Street entrance squares, south and north Roman Street inner plazas and the waterside plaza near Wangju Lake in the east.

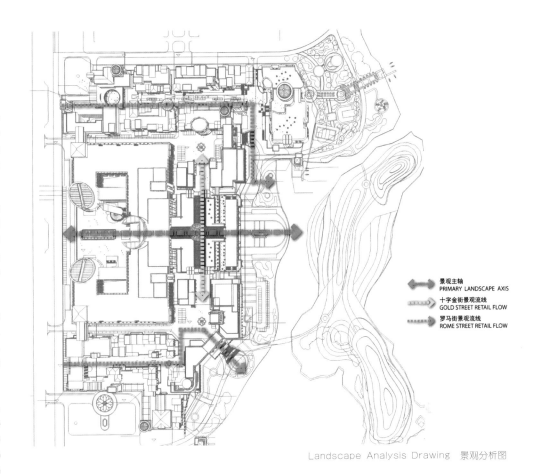

Landscape Analysis Drawing　景观分析图

■ 规划理念

项目的规划理念为"一轴、一心、双环、六节点":

"一轴"——由沿长宁大道主广场—奥特莱斯主题建筑—十字金街—滨水广场—王咀湖构成东西方向的主轴线,既是主要的商业动线,也是景观轴线。

"一心"——由奥特莱斯主体建筑与十字金街组成的核心建筑群,作为整个项目的重点空间,聚集了国际各大主品牌,是整个项目的中心节点和精神空间。

"双环"——一环由奥特莱斯主体建筑与周围罗马风情街形成,为基础内部主要机动车流线。另一环由罗马风情街本身形成,为主要购物步行街流线。

"六节点"——由奥特莱斯主体建筑前靠近长宁大道一侧广场、南北罗马街入口广场、南北罗马街内广场及东侧临王咀湖滨水广场构成六个景观空间节点。

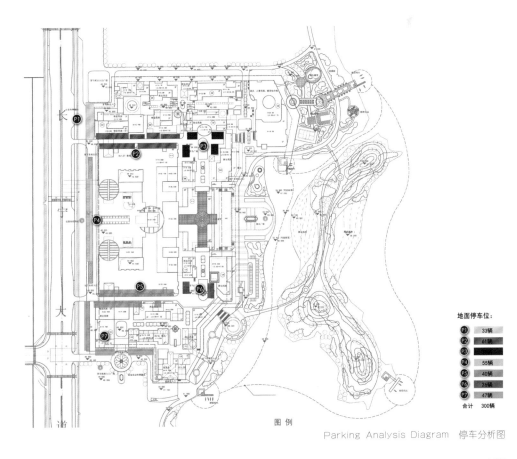

Parking Analysis Diagram　停车分析图

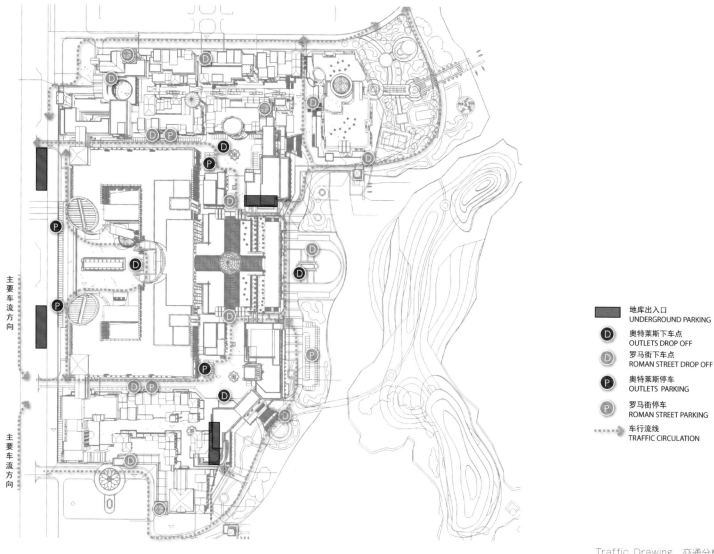

Traffic Drawing 交通分析图

■ Functional Layout

For the functional layout, the main Outlets building is a multi-storey building with four layers above the ground and one layer underground and its architectural form is set in axial symmetry. The main function of the underground layer is served for retail and restaurant, and linked with the underground garage. Four layers above the ground set clothing discount pavilion and cinema, etc. The Cross Gold Street is a terrace-backwards building along Wangju Lake with four layers above the ground and one layer underground. The layer underground sets about 3,000 m² boutique supermarket, , restaurant and bar, and the first and second layer above the ground are mainly Italian boutiques clothing hall, brand jewelry and other high-end shops, and the third and forth layers are functioned for food and beverage, fully enjoying the beautiful scenery of Wangju Lake.

The Roman Street is divided into south and north two blocks, and respectively sets retail shops, restaurants, KTV and other entertainment forms, etc. The inner part of Roman Street adopts design of water landscape combined with shops on both sides and three layers of terraces, forming rich shopping and leisure space.

The complex building with functions of wedding, adventure playground and fitness gym is a three to four layers architecture that the first and second layers are for the wedding, the third layer is for the children's adventure playground and the forth layer is a fitness center. The design is combined with the outdoor landscape to decorate the wedding lawn, children paddling area and fitness field, etc.

■ 功能布局

在功能布局上，主体奥特莱斯为地上四层、地下一层的多层建筑，建筑形体按凹字形中轴对称布置。地下一层的主要功能为零售及餐饮，并与地下车库相联。地上一至四层设置服装折扣馆和影院等。十字金街为地上四层、地下一层，沿王咀湖逐层退台式的建筑。地下设置3 000 m²左右的精品超市及酒吧餐厅，地上一至二层布置意大利精品服饰馆、品牌珠宝馆及其他高端主力店，三层至四层为餐饮功能，尽享王咀湖美景。

而罗马街分为南北两区，分别布置零售商铺、餐饮、KTV及其他娱乐业态等。罗马街内街采用景观水街的设计，结合两边商铺及三层露台，形成空间丰富的购物休闲空间。

婚庆儿童乐园健身综合楼为三至四层建筑，一、二层为婚庆庄园，三层为儿童乐园，四层为健身中心。设计结合室外景观布置婚庆草坪、儿童戏水区和健身场地等。

Elevation 1 立面图 1

Elevation 2 立面图 2

Elevation 3　立面图 3

Elevation 4　立面图 4

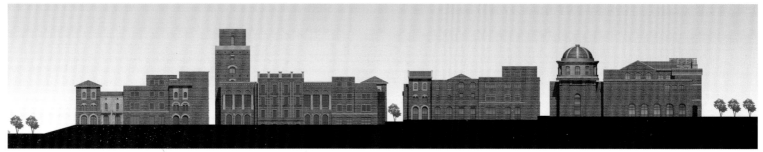
Elevation 5　立面图 5

Elevation 6　立面图 6

Elevation 7　立面图 7

Elevation 8　立面图 8

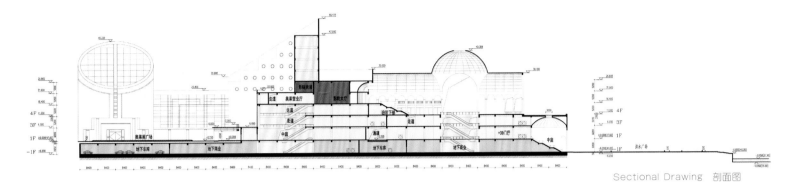

Sectional Drawing 剖面图

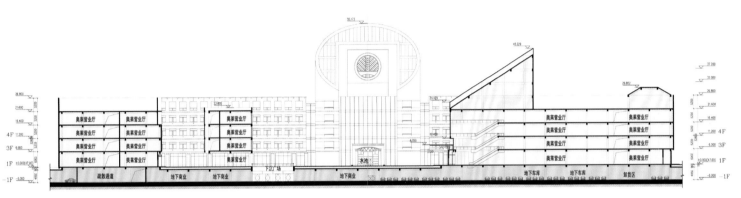

Sectional Drawing 剖面图

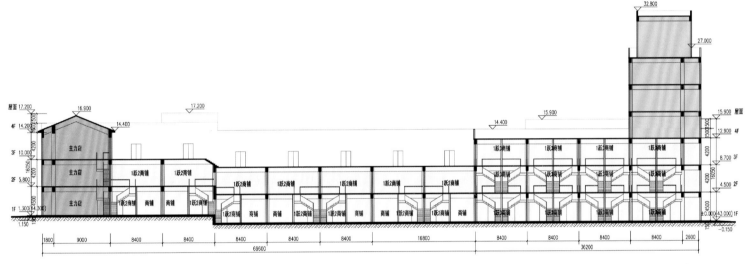

Section 1-1 剖面图 1-1

■ Building Facade Design

For the building facade, the main building Outlets adopts the Swiss architect Mario Botta's architectural style that it makes wonderful extraction on the traditional European architectural language and applies facade processing technique of tile collage, which looks concise with rich details, and through the comparing of the scale and materials, it highlights the texture sense and taste of the building. The architectural style and space intention of Cross Gold Street directly originate from Milan Vittorio Emanuele II arcade, using super-scale arcade and dome to create grand commercial space node, thus making it become the spirit core of the whole project. The Roman Street adopts small town residential architectural style of ancient Rome. Most buildings take 2 or 3 layer residence style. The roof contour line goes orderly up and down, forming rich skyline; at the same time, it makes subtle contrast with the main Outlets building. The street space has modest density and pleasant scale, which forms a friendly neighborhood space and provides a perfect and comfortable environment for shopping.

■ 建筑立面设计

在建筑立面上，主体奥特莱斯采用瑞士建筑大师马里奥·博塔的建筑风格，对传统的欧式建筑语言进行精彩的提炼，采用面砖拼贴的立面处理手法，简洁中富含丰富的细节，通过尺度、材料的对比，突显建筑质感与品位。而十字金街的建筑风格及空间意向则直接取材于米兰埃马努埃莱二世拱廊，以超尺度的拱廊及穹顶构筑了恢弘的商业空间节点，成为整个项目的精神核心。罗马街采用了古罗马小镇民居式的建筑风格。建筑多为2~3层民居式风格。屋顶轮廓线高低起伏、错落有致，形成丰富天际线的同时，又与奥特莱斯主体建筑形成微妙的反差和对比。街道空间疏密有致，尺度宜人，形成亲切的邻里空间，为购物提供良好舒适的环境。

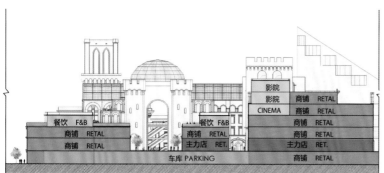

Section Sketch 1 剖面示意图 1

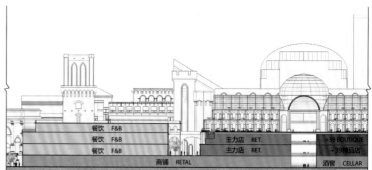

Section Sketch 2 剖面示意图 2

Second Floor Plan 二层平面图

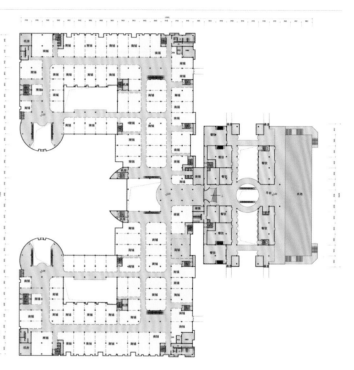

Third Floor Plan 三层平面图

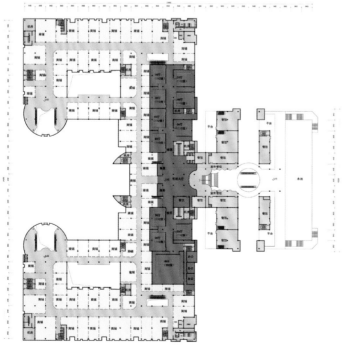

Fourth Floor Plan 四层平面图

Fifth Floor Plan 五层平面图

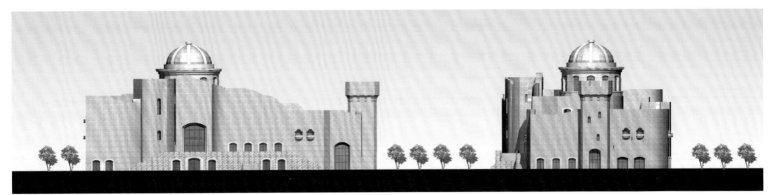

Elevation 9　立面图 9

Elevation 10　立面图 10

023

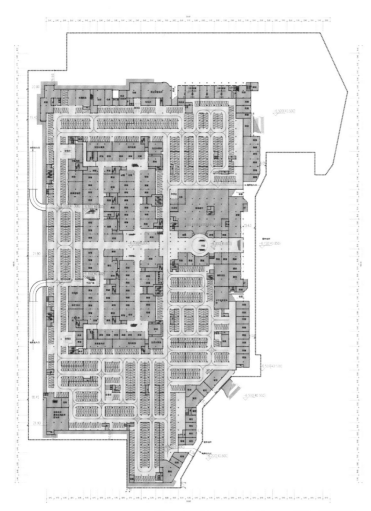

Plan for Basement Floor 地下一层平面图

Piece Plan 1st Floor 拼合平面图 1 楼

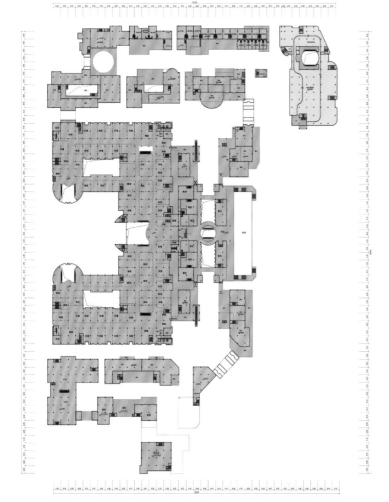

Piece Plan 3rd Floor 拼合平面图 3 楼

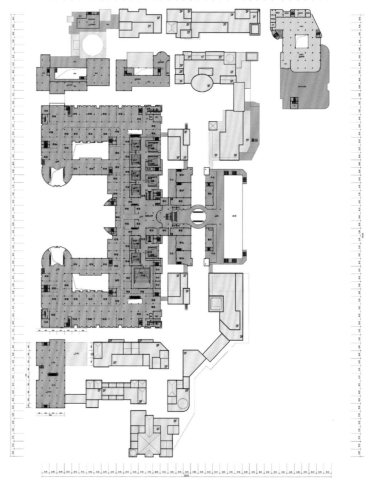

Piece Plan 4th Floor 拼合平面图 4 楼

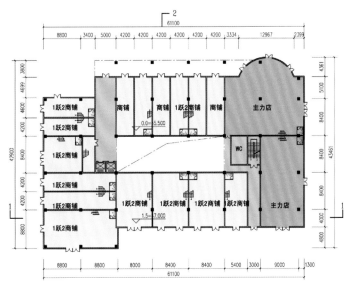
First Floor Plan 一层平面图

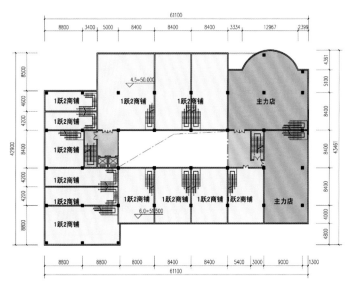
Second Floor Plan 二层平面图

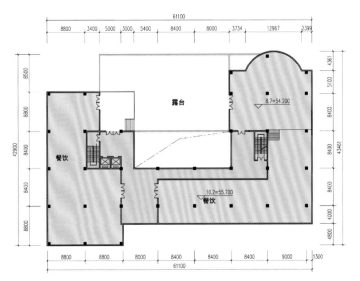
Third Floor Plan 三层平面图

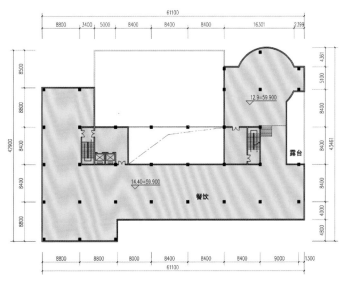
Fourth Floor Plan 四层平面图

Yibin NHH Fortune Center
宜宾唐人财富中心

Keywords 关键词

Axial Symmetric
中轴对称

Fashionable Element
时尚元素

Grand Modeling
造型大气

Location: Yibin, Sichuan, China
Architectural Design: China Construction International (CCI) Architecture Design & Consulting Co., Ltd.
Net Planning Site Area: 56,494.35 m²
Total Floor Area: 175,715 m²

项目地点：中国四川省宜宾市
设计单位：上海新外建工程设计与顾问有限公司
规划净用地面积：56 494.35 m²
总建筑面积：175 715 m²

Features 项目亮点

The facade style adopts eclecticism technique, which not only reflects the high grade of the classic business, but also blends in the fashionable elements of modern business. Stone materials, metal and glass are organically constituted to put emphasis on the vertical sense.

立面风格采用折中主义手法，既体现经典商业的高档，又融入了现代商业的时尚元素。石材、金属、玻璃通过有机构成，强调纵向感。

■ **Overview**

The project is located at the junction of Yibin Jinshajiang Avenue and Chongwen Road; it has convenient transport, complete life supporting facilities and residential community around, but lacks commercial complex. The project includes high-grade medium and high-rise residential complexes, of which buildings 4 and 5 are 32-layer residences with bottom-layer business use, buildings 2 and 3 are for the store business and building 1 is independent comprehensive business area.

■ 项目概况

项目位于宜宾金沙江大道和崇文路交接处，交通便利，生活配套齐全，周边大多为住宅小区，欠缺商业综合体。项目为包含高档中高层住宅区的综合体，其中4、5号楼为32层低层商用住宅，2、3号楼为店铺式商业，1号楼为独立综合商业区。

■ **Planning and Design**

The regional planning and design can be summarized as "one center, one ring and four nodes". "One center" refers to the complex building including supermarket, cinema, Gome and large catering space as a whole. "One ring" refers to the commercial pedestrian street, which is a commercial block bearing brands, leisure, entertainment and shopping. "Four nodes" refers to the four squares, which is an urban development space for people gathering and integrates entertainment, leisure and landscape in whole.

■ 规划设计

本地域规划设计为"一心一环四节点"。"一心"指集超市、电影院、国美、大型餐饮于一体的综合楼。"一环"指商业步行街，它是集品牌、休闲、娱乐、购物为一身的商业街区。"四节点"则是指其中的四个广场，作为人流汇聚场所，是集娱乐、休闲、景观为一体的城市开发空间。

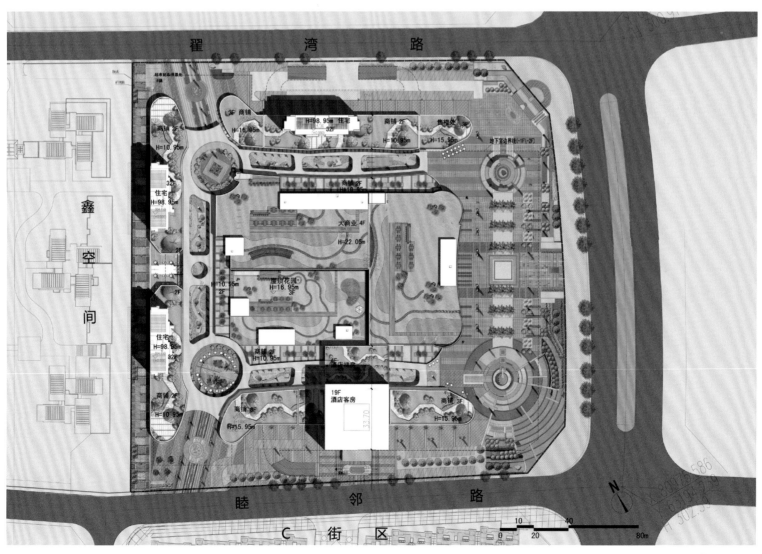

Regulation Plan 规划总平面图

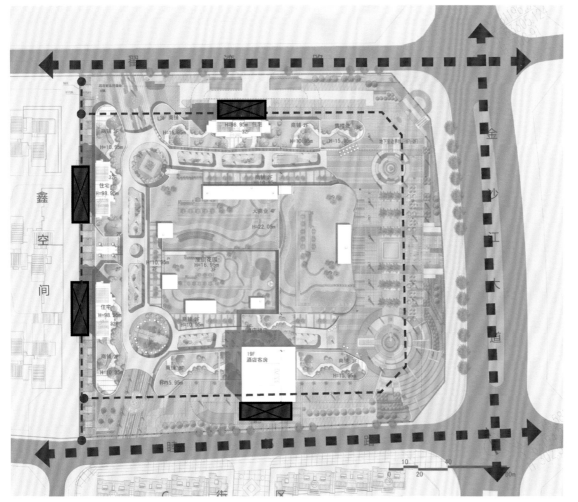
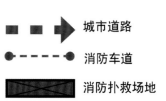

Fire Analysis Diagram 消防分析图

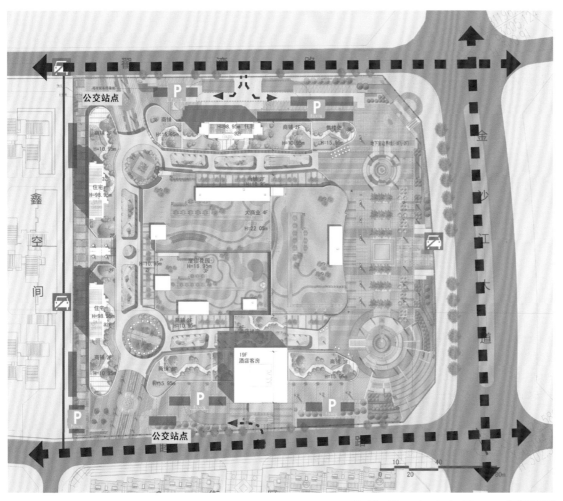
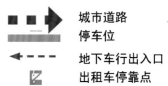

Traffic Flow Analysis Diagram 车流分析图

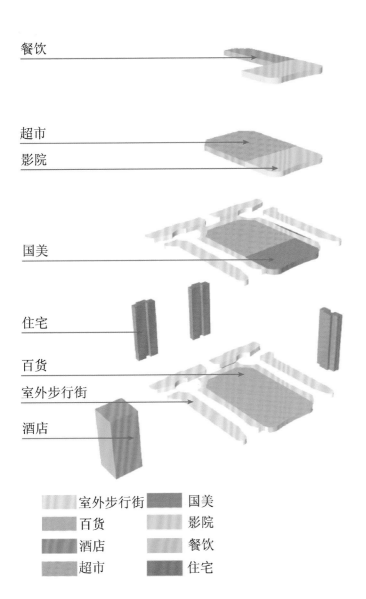

- 餐饮
- 超市
- 影院
- 国美
- 住宅
- 百货
- 室外步行街
- 酒店

室外步行街	国美
百货	影院
酒店	餐饮
超市	住宅

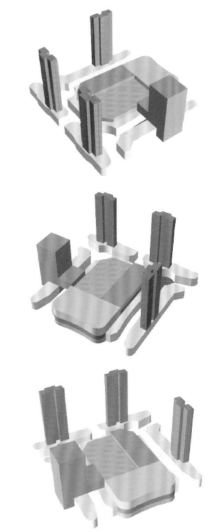

Functional Distribution Analysis Drawing
功能分布分析图

■ Traffic Organization Design

The traffic organization is based on the principle of the original fluent but jammed roads to create an order and clear functional road organization, and meet the requirements of fire control and ambulance rescue as well. The vehicle transportation organized by circular main trunk roads is combined with the green environment and space layout to form perfect road landscape, and makes the central landscape walking system relatively independent. At the same time, the road design is cooperated with the square space, green space, building space together to shape the outdoor landscape, achieving the perfect combination of function and form.

■ 交通组织设计

在交通组织方面，整体依据居住区道路通而不畅的原则，形成分级有序，功能明确的道路组织，同时满足消防、救护等要求。以环状的主干道路组织功能组团之间的车行交通，结合绿化环境与空间布局，形成良好的道路对景效果，并使中心景观步行系统相对独立。同时，道路设计与广场空间、绿地空间、建筑空间相结合，共同塑造户外空间景观，实现功能与形式的完美结合。

■ Construction and Facade Design

In order to reflect the first city commercial comprehensive flagship brand in Yibin, the overall architectural modeling looks lively and grand, and the space layout takes axial symmetry technique. The facade style adopts eclecticism technique, which not only reflects the high grade of the classic business, but also blends in the fashionable elements of modern business. Stone materials, metal and glass are organically constituted to put emphasis on the vertical sense. Each residential building has a barrier-free ramp into the lobby. Each layer is convenient for the wheelchair, and the elevator room satisfies the requirement of wheelchair turning around.

■ 建造与立面设计

为体现打造宜宾第一城市商业综合旗舰品牌，整体建筑造型明快大气，空间布局中轴对称。立面风格采用折中主义手法，既体现经典商业的高档，又融入了现代商业的时尚元素。石材、金属、玻璃通过有机构成，强调纵向感。各幢住宅都设有无障碍的坡道进入大堂，各层建筑便于轮椅到达，电梯前室满足轮椅回旋要求。

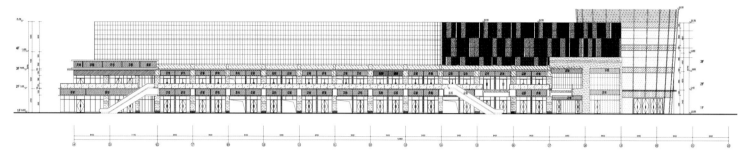

Side Elevation 1　轴立面图1

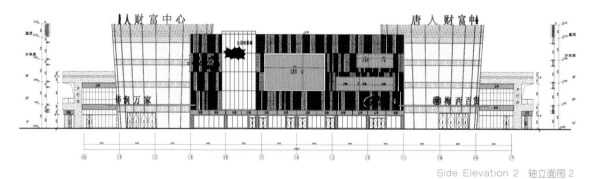

Side Elevation 2　轴立面图2

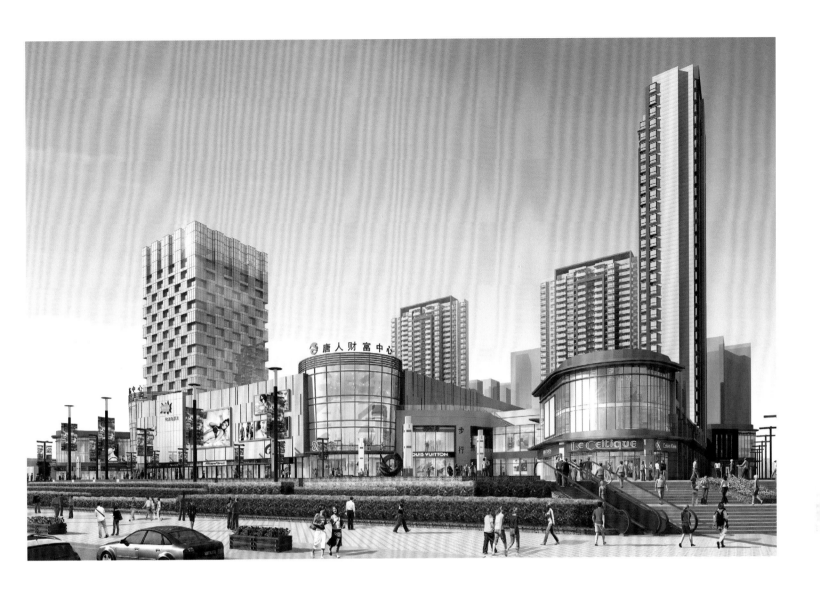

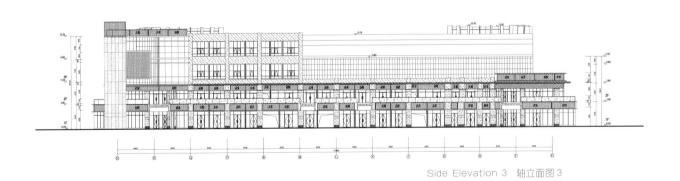

Side Elevation 3 轴立面图 3

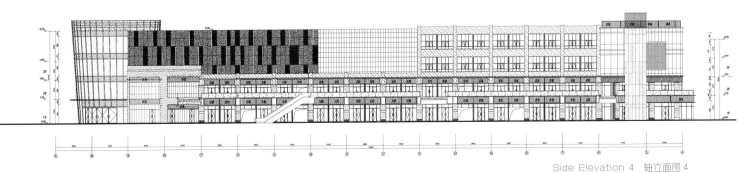

Side Elevation 4 轴立面图 4

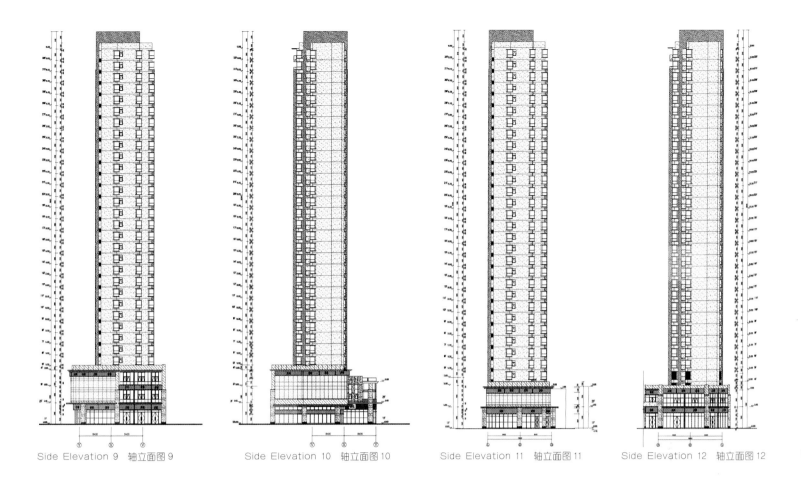

Side Elevation 9 轴立面图9　　Side Elevation 10 轴立面图10　　Side Elevation 11 轴立面图11　　Side Elevation 12 轴立面图12

Side Elevation 5 轴立面图 5

Side Elevation 6 轴立面图 6

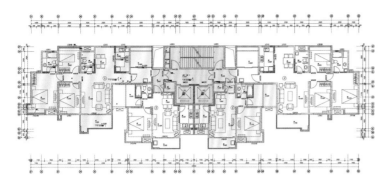

Residence Block 1 Standard Layer Plan 住宅一号楼标准层平面图

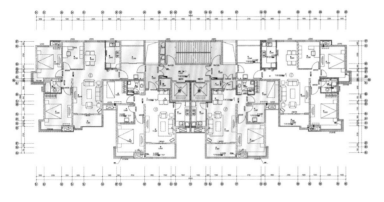

Residence Block 2 Standard Layer Plan 住宅二号楼标准层平面图

Shenyang Chanceave City Mall
沈阳千姿汇购物中心

Keywords 关键词

Unique Appearance
外观独特

Bright-colored
色彩斑斓

People-oriented
以人为本

Location: Shenyang, Liaoning, China
Architectural Design: SJAT Architecture & Engineering Design
Total Site Area: 9,657.06 m²
Total Floor Area: 44,700 m²
Overground Floor Area: 23,000 m²
Underground Floor Area: 21,700 m²
Plot Ratio: 2.4

项目地点：中国辽宁省沈阳市
设计单位：北京世纪安泰建筑工程设计有限公司
总用地面积：9 657.06 m²
总建筑面积：44 700 m²
地上建筑面积：23 000 m²
地下建筑面积：21 700 m²
容 积 率：2.4

Features 项目亮点

In accordance with people-oriented principles, designers pay close attention to customers' psychology and behavior, creating cultural, public, comfortable, humanized commercial space, thus to provide an open indoor shopping and leisure space environment.

项目设计坚持"以人为本"的设计原则，密切关注顾客的心理和行为，关注文化性、公共性、舒适性、人性化的商业空间的塑造，创造富于人性化的、开放的室内购物休闲空间环境。

■ **Overview**

The project's convenient location creates favorable condition for commercial activities. The first floor underground, first floor and second floor are dedicated to commerce, the third floor is for commerce and catering, the fourth floor is cinema, the fifth floor is for cinema subsidiary rooms, the third floor underground and second floor underground are parking space.

■ **项目概况**

项目地段交通极为便利，为商业活动创造了有利的条件。地下一层、地上首层与二层为商业，三层为商业与餐饮，四层为影院，五层为影院辅助用房，地下三层与地下二层为车库。

■ **Architectural Design**

As a commercial complex, Shenyang Chanceave City Mall takes fully account of urban design principles, becoming a bright spot among the building group in this region and adding splendor to the construction of Shenyang Tiexi. In accordance with people-oriented principles, designers pay close attention to customers' psychology and behavior, creating cultural, public, comfortable, humanized commercial space, thus to provide an open indoor shopping and leisure space environment. Modern advanced technology on electrical network and advertising display is used to offer customers convenient traffic and service, efficient organization and management, favorable security environment and passionate advertising display. In addition, natural environment is introduced into interior shopping space. Outdoor commercial plaza and north roof space are dedicated to creating multi-level commercial environment.

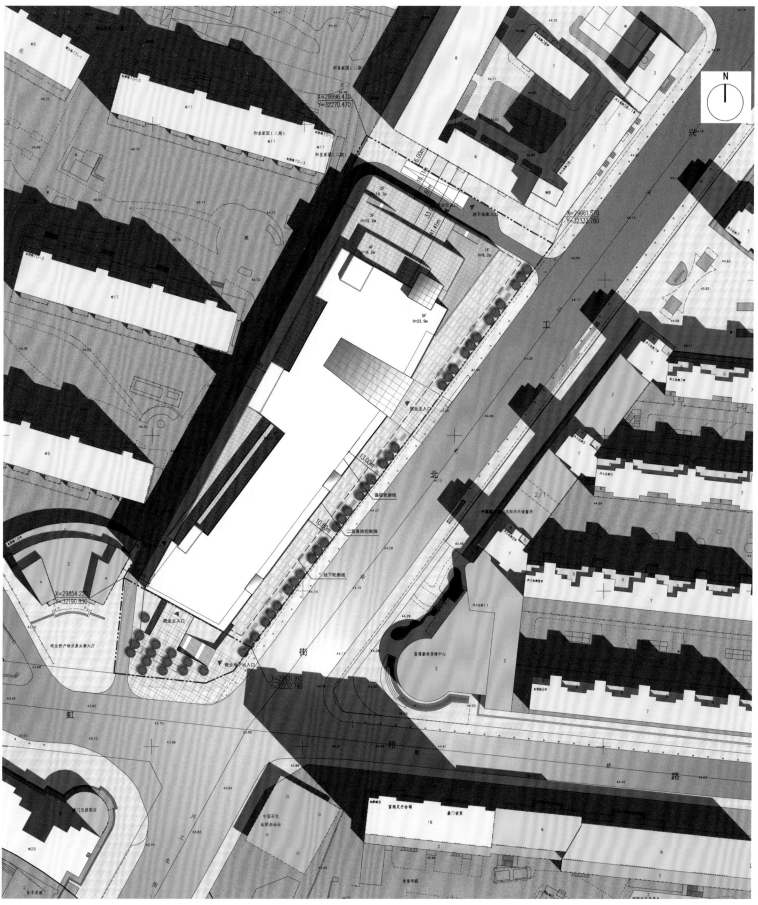

Site Plan 总平面图

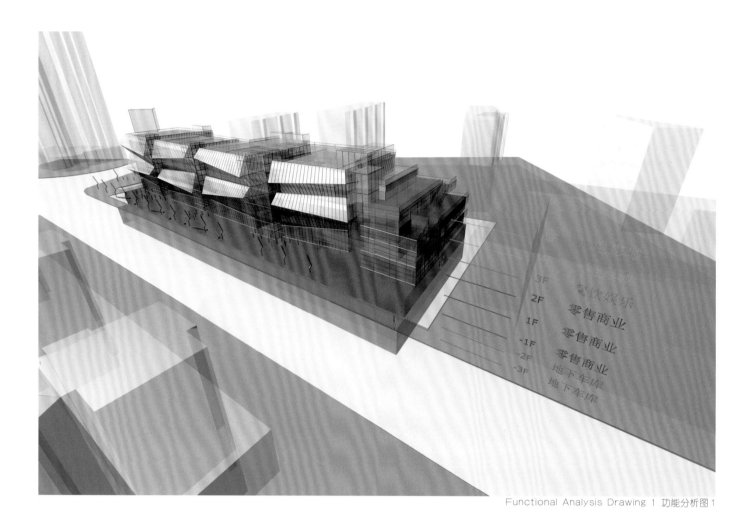

Functional Analysis Drawing 1 功能分析图1

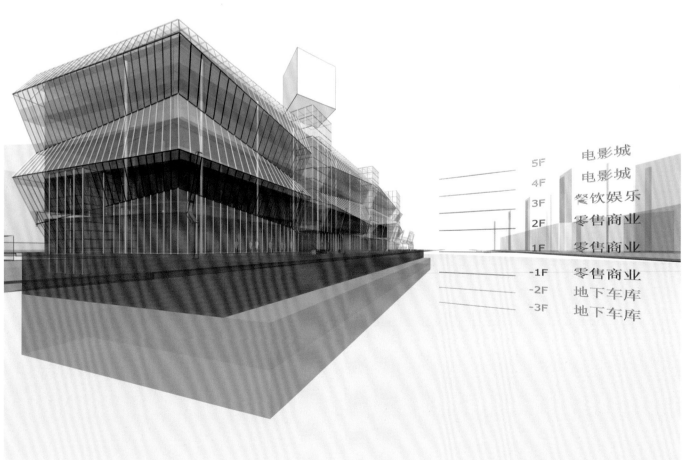

Functional Analysis Drawing 2 功能分析图2

■ 建筑设计

　　沈阳千姿汇购物中心作为商业综合体充分考虑城市设计的原则，成为该区域建筑群中的一个亮点，为沈阳市铁西区的建设增添光彩。设计坚持"以人为本"的设计原则，密切关注顾客的心理和行为，关注文化性、公共性、舒适性、人性化的商业空间的塑造，创造富于人性化的、开放的室内外购物休闲空间环境。设计利用现代先进的电子网络、广告展示等科学技术，为顾客提供便捷的交通、服务，有效的组织、管理，良好的治安环境和富于激情的广告展示。另外，设计还把自然环境充分引入室内购物空间、室外商业广场和北侧屋顶空间创造多层次的商业环境。

Star City International Plaza
城际星港城

Keywords 关键词

Glass Curtain Wall
玻璃幕墙

Aluminum Sheet Curtain Wall
铝板幕墙

LED Lighting
LED 灯饰

Features 项目亮点

To achieve a balance between commercial practicality and architectural aesthetics, a large area of glass curtain wall and aluminium sheet curtain wall are used to form a sense of rhythm, and LED lighting is used to achieve a strong commercial effect.

本商业广场力求达到商业实用性与建筑美学的平衡，采用了大面积的玻璃幕墙与铝板幕墙，利用不同材质形成有节奏的韵律感，大范围地利用LED灯饰给商场带来强大的商业效应。

Location: Foshan, Guangdong, China
Developer: Intercity Real Estate
Planning, Architectural Design: AIM International
Architect: Chen Xiaoyu
Total Land Area: 25,368 m²
Total Floor Area: 310,000 m²
Plot Ratio: 5.0

项目地点：中国广东省佛山市
开 发 商：城际置业
规划/建筑设计：加拿大AIM国际设计集团
主创设计：陈晓宇
总占地面积：25 368 m²
总建筑面积：310 000 m²
容 积 率：5.0

■ Design Concept

The project is conceived to shape a future urban commercial complex which includes three-dimensional city, city information station, comprehensive life network and green ecological buildings.

■ 设计概念

本方案规划构想以塑造未来城市商业体为原型，设计概念包括：立体城市、都会信息站、全方位生活网络、绿色生态建筑。

■ Architectural Design

To achieve a balance between commercial practicality and architectural aesthetics, a large area of glass curtain wall and aluminium sheet curtain wall are used to form a sense of rhythm, and LED lighting is used to achieve a strong commercial effect, which bring an emerging cultural scene with modern design concept for Jinshazhou and surrounding area.

■ 建筑设计

本商业广场力求达到商业实用性与建筑美学的平衡，采用了大面积的玻璃幕墙与铝板幕墙，利用不同材质形成有节奏的韵律感，大范围地利用LED灯饰给商场带来强大的商业效应，使商场为金沙洲乃至周边地区带来一片具有现代设计理念的新兴文化景象。

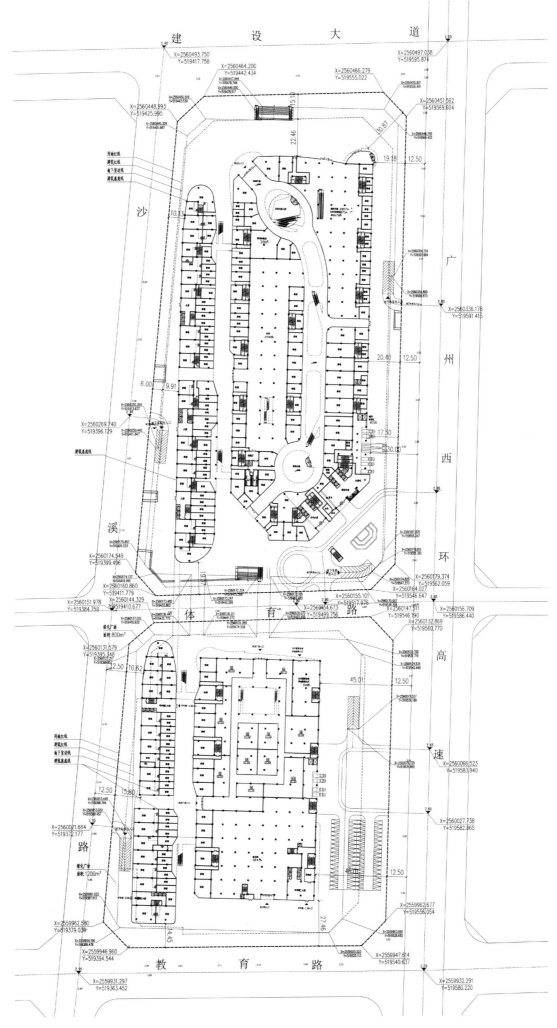

Site Plan 总平面图

045

Penglai Joy Plaza
蓬莱悦动港湾购物街区

Keywords 关键词

One-stop
一站式

Cultural Atmosphere
文化氛围

Function Division
功能分区

Location: Yantai, Shandong, China
Architectural Design: Haskoll (Beijing) Architectural Design Consultancy
Developer: Beijing White Peak Real Estate
Area: 150,000 m²

项目地点：中国山东省烟台市
设 计 单 位：北京赫斯科建筑设计咨询有限公司
开 发 商：北京鼎峰地产投资顾问有限公司
面　　　积：150 000 m²

Features 项目亮点

Main attention was paid on developing entertained & recreational programs in this design, and on this basis more elements were added to heighten profound cultural atmosphere. In terms of function division, retail area and catering area were separated to attract more consumers to have fun in this wonderful plaza.

该项目以开发休闲娱乐活动为主，在此基础上加入了更多的元素来烘托浓厚的文化氛围。在功能分区上把零售区域、餐饮区域分隔开来，目的是为了吸引更多的消费者到此来休闲和娱乐。

■ Overview

This project is located between the main seaside tourist attraction and CBD in Penglai, Shandong. Designers proposed to connect the two areas in terms of overall planning thanks to its important geographical location. In addition, standing on the famous Penglai Pavilion, one can catch a panoramic view of this complex clearly.

■ 街区概况

　　该项目位于山东省蓬莱市海滨主要旅游景点与城市中央商业区之间，由于该项目在城市中占据重要的地理位置，设计公司在整体规划设计上提出连接两个区域的规划设想。此外，站在著名的蓬莱阁上能清晰地俯瞰到这一综合体的全景。

■ Street Design

Main attention was paid on developing entertained & recreational programs in this design, as shown in the project name "Joy Plaza". Besides movie theatre and KTV, there is a four-storey business and leisure hotel. And on this basis more elements were added to heighten profound cultural atmosphere, such as culinary school and red wine experience center. In terms of function division, retail area and catering area were separated to attract more consumers to have fun in this wonderful plaza.

■ 街区设计

　　该项目被命名为"悦动港湾"，强调以开发休闲娱乐活动为主，不但包括电影院、KTV等场所设施，还配有4层楼的休闲商务酒店。设计师还在此基础之上加入了更多的元素来烘托浓厚的文化氛围，例如烹饪学校和红酒体验中心。在功能分区上把零售区域、餐饮区域分隔开来，目的是为了吸引更多的消费者到此来休闲和娱乐。

Site Plan 总平面图

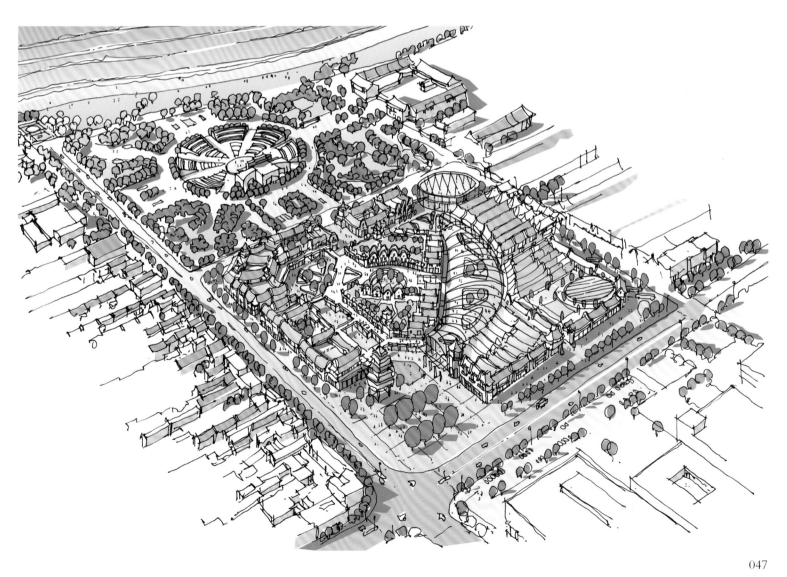

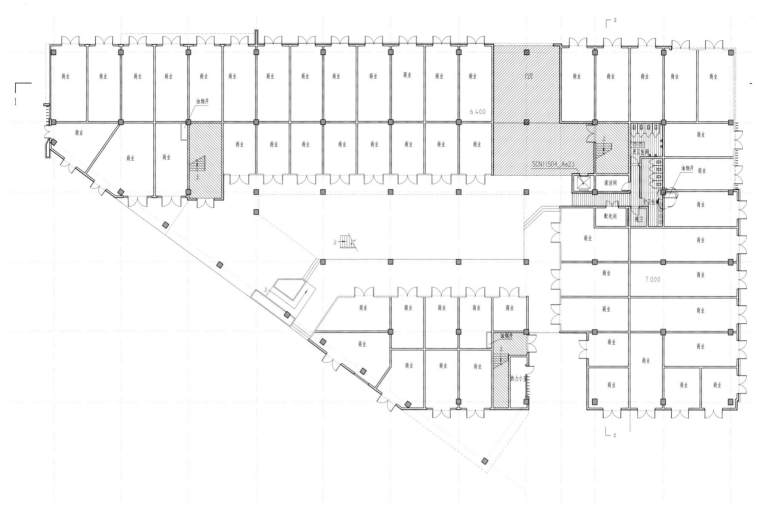

Ground Level Floor Plan 首层平面图

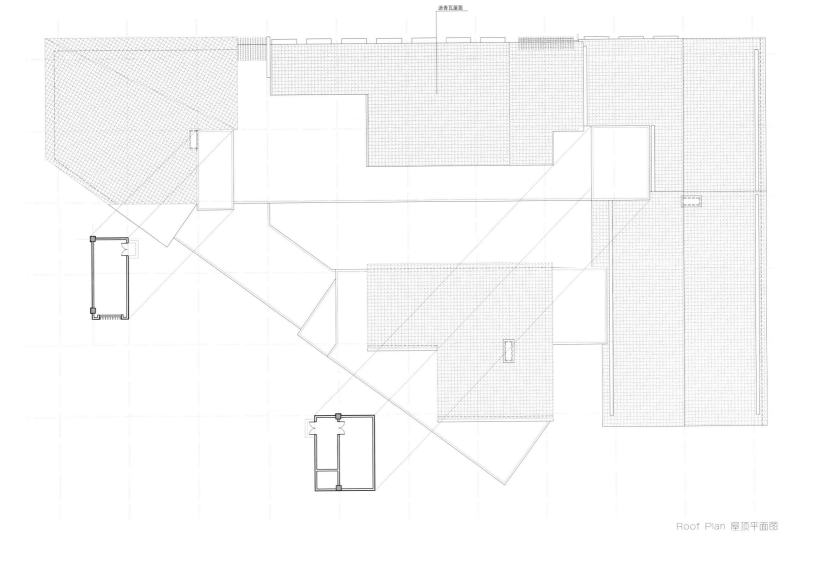

Roof Plan 屋顶平面图

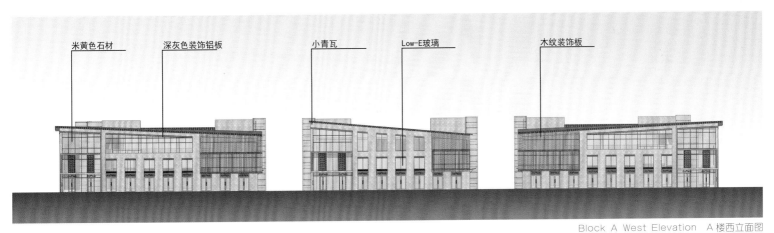

Block A West Elevation　A楼西立面图

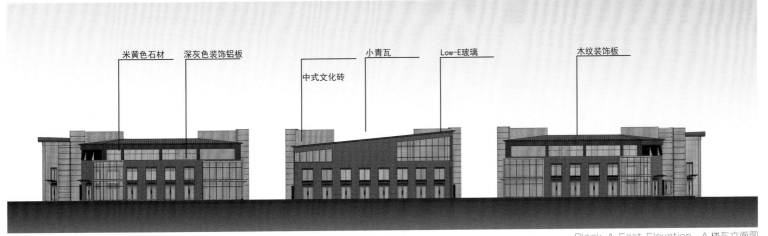

Block A East Elevation　A楼东立面图

Side Elevation 轴立面图

Side Elevation 轴立面图

Side Elevation 轴立面图

Side Elevation 轴立面图

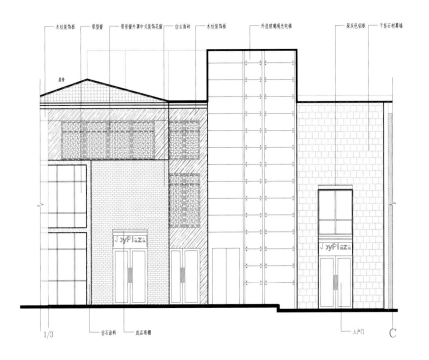
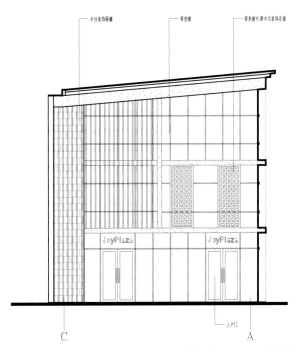

Side Elevation 轴立面图

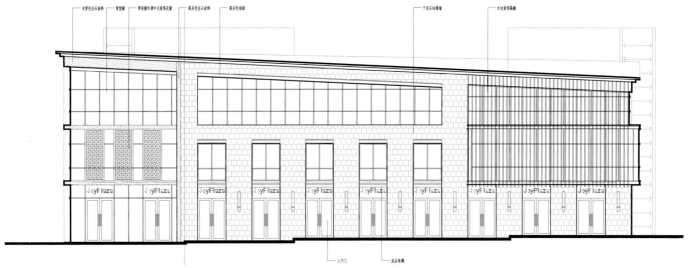

Side Elevation 轴立面图

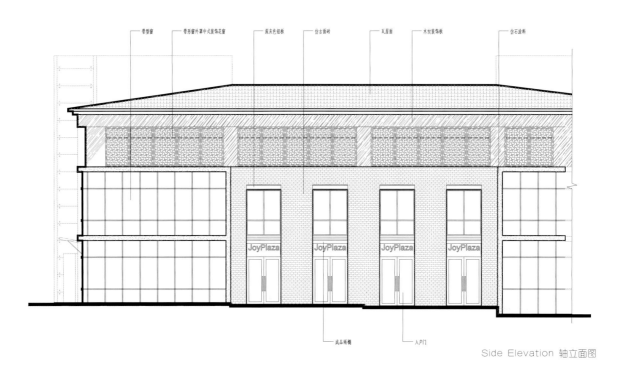

Side Elevation 轴立面图

Side Elevation 轴立面图

Side Elevation 轴立面图

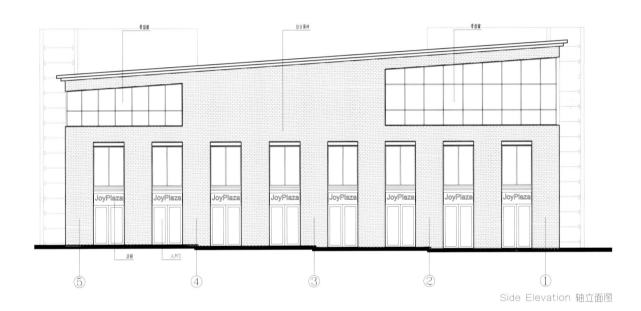

Side Elevation 轴立面图

Commercial Plaza on Nanjing Road, Tianjin
天津南京路商业广场

Keywords 关键词

- Attractive Atrium 魅力中庭
- Unique Style 独特风格
- Wayfinding System 导视系统

Location: Heping District, Tianjin, China
Architectural Design: KAZIA.LI Design Collaborative

项目地点：中国天津市和平区
设计单位：天津凯佳李建筑设计事务所

Features 项目亮点

Based on a set of design guidelines, the architects leave the freedom to the merchants to decorate their shops, thus to create a colorful and exciting indoor environment.

在遵循一系列设计指引的基础上，项目给予商家发挥的最大空间，使商家可以任意设计自己的店面形象，从而满足购物中心丰富而绚丽的内部效果。

■ Atrium Space

The research of the focus group indicated that the public do not like the enclosed and internalized shopping environment, but prefer that with access to the sunshine and connecting with the outdoor, corresponding with the unique qualities of the city or town. For a shopping plaza, it is quite important to have an attractive atrium to let people willing to stay.

■ 商业中庭

焦点小组的研究表明公众不喜欢封闭化的内在式购物环境，喜欢与室外环境相通，能感受阳光的自然环境，并与城市或城镇的独特气质相符。对于购物中心来说，一个好的中庭设计必须要引人注目，能够吸引人们，让人们乐意在这个室内闲逛。

■ Style and Features

Every shopping mall has its own style and characteristics, and a unified aesthetic design can express the overall integrity of the mall brand. Based on a set of design guidelines, the architects leave the freedom to the merchants to decorate their shops, thus to create a colorful and exciting indoor environment.

■ 风格特点

每个购物中心都具有它自己的风格和特点，基调的整体统一能更好地反映购物中心的完整性。在此基础上，给予商家发挥的最大空间，使商家可以任意设计自己的店面形象，从而满足购物中心丰富而绚丽的内部效果。最终达到整体而不呆板，丰富而不混乱的效果。

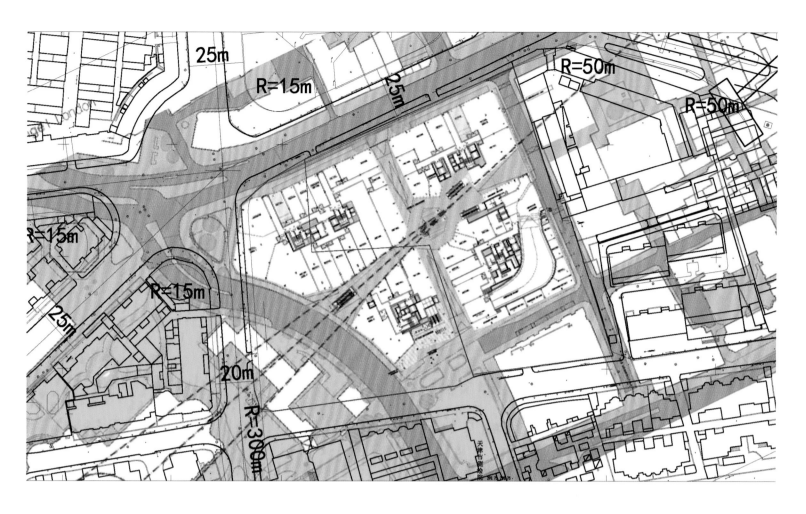

■ Way Finding Design

A successful shopping center must have excellent way finding and branding design. Effective way finding design can inform appropriate placement of goods and services, and have great influence on the interior design.Thus, it should be coordinated with the branding of the overall environment. The way finding design should not only be beautiful but also be functional.

■ 导视系统设计

一个成功的购物中心绝对离不开一套成功的导视系统设计和形象VI设计。导视系统设计能提供购物中心卓越的商品导购力与道路指引力，它对室内设计的最终效果影响很大，所以导视系统设计和形象VI设计应做到跟整个环境完美地融合。导视系统的设计不仅应从美观上考虑，还应该多从功能设计并且满足人性化的要求上考虑。

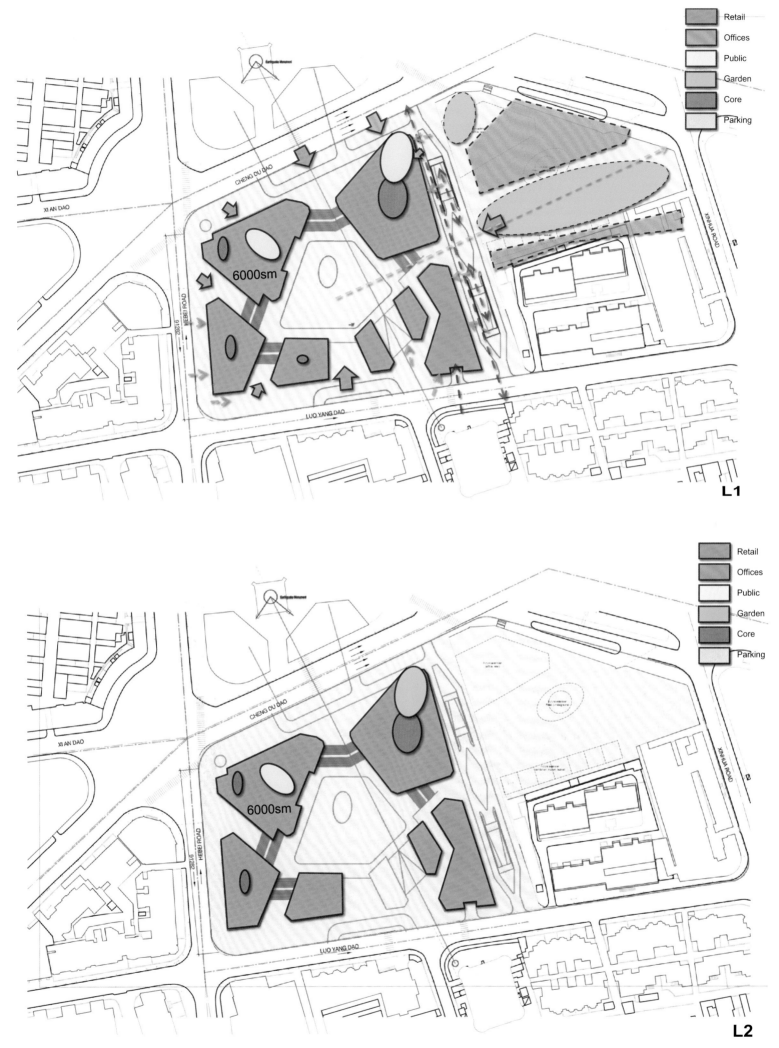

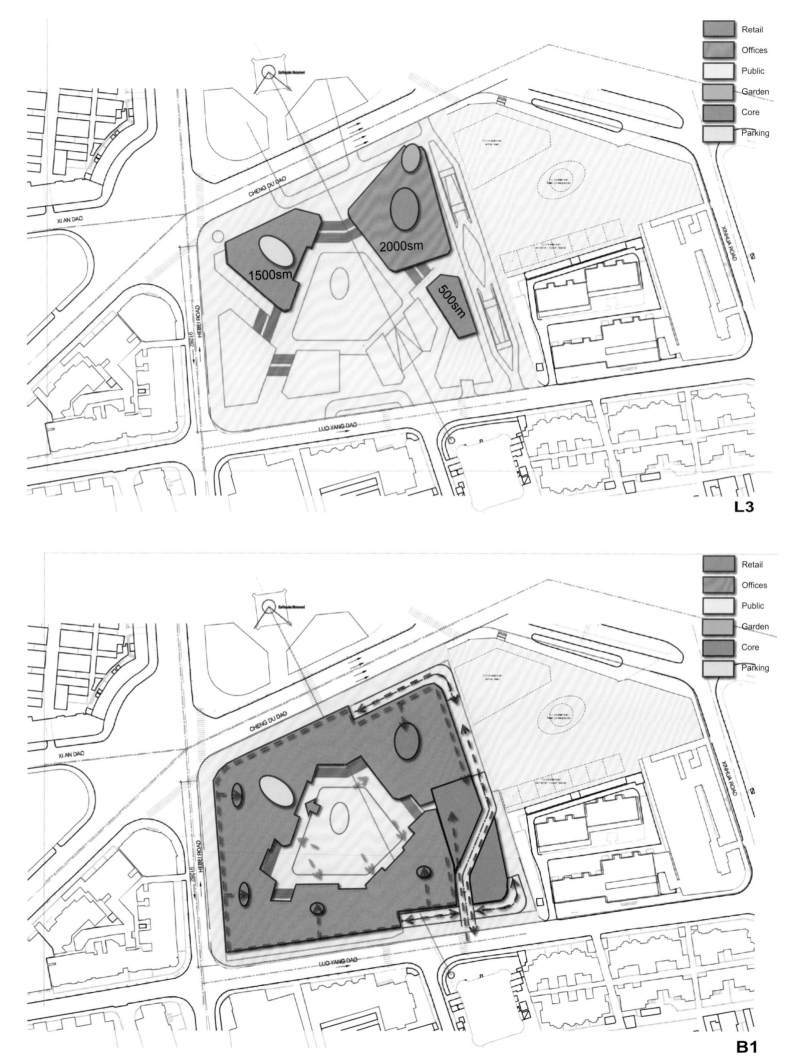

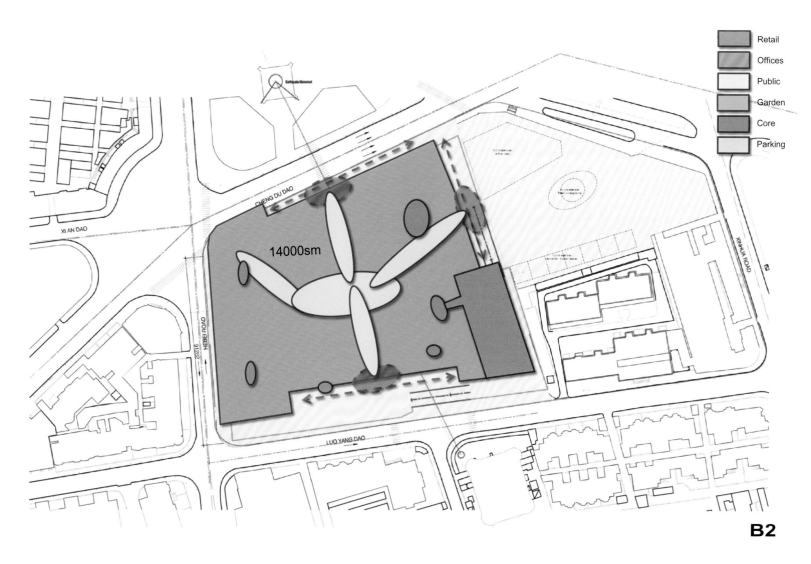

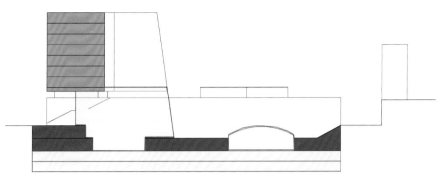

Winter Gardens and Sunken Plaza

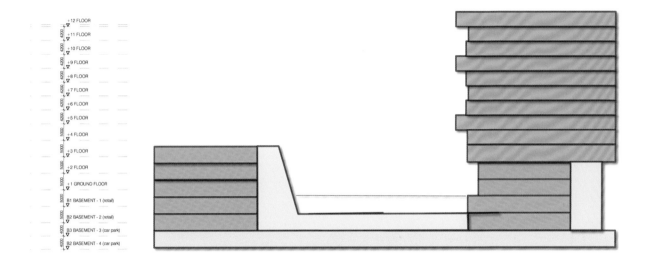

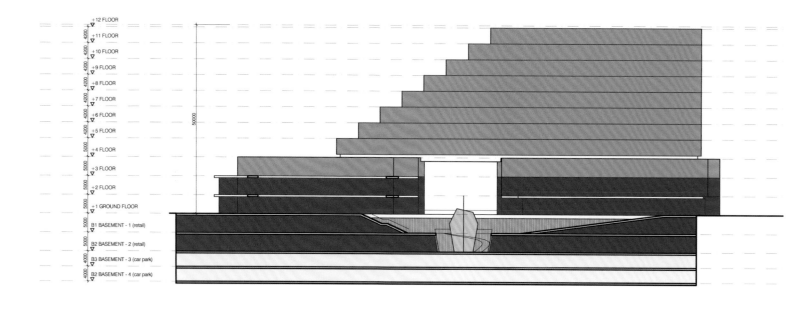

Fengdong International
大明宫·沣东国际

Keywords 关键词

- **Unified Style** 风格统一
- **Material Effect** 材料效应
- **Generous Posture** 姿态大气

Features 项目亮点

The building is in simple modern style, which pays attention to the details of facade and highlights the unification of architectural form, style and proportion. Grand Shape as complete street front and large plaza are used to interpret the relationship between the building and the city interface.

建筑以简约现代风格为主，在立面造型设计上注重细节的表达，强调建筑形式、风格、比例的统一塑造。项目以完整的沿街面、大型的广场等大气姿态，来诠释建筑与城市界面的关系。

Location: Xi'an, Shaanxi, China
Developer: Xi'an Qinfeng Investment Development Co., Ltd.
Architectural Design: Sichuan Cendes Architecture Engineering Design Co., Ltd.
Land Area: 18,612.79 m²
Floor Area: 244,047.13 m²
Plot Ratio: 3.99
Green Coverage Ratio: 15%

项目地点：中国陕西省西安市
开 发 商：西安秦沣投资发展股份有限公司
建筑设计：四川山鼎建筑工程设计股份有限公司
占地面积：18 612.79 m²
建筑面积：244 047.13 m²
容 积 率：3.99
绿 化 率：15%

■ Overview

The project is located in Fengdong New Town, Xixian New Area, Xi'an, which is a core section dedicated to create international metropolis. The site boarders Houweizhai Interchange urban green space to the west, while Sanqiao Sub-district Office to the east, Sanqiao Central Street to the south and urban expressway viaduct to the north.

■ 区位概况

项目位于西安西咸新区沣东新城，为打造国际化大都市的核心地段。用地西起后围寨立交城市规划绿地，东至三桥街办现址，南起三桥正街，北至城市快速高架桥。

■ Architectural Design

The building is in simple modern style, which pays attention to the details of facade and highlights the unification of architectural form, style and proportion. Grand Shape as complete street front and large plaza are used to interpret the relationship between the building and the city interface. In addition, attention is also paid to materials, i.e., selecting appropriate materials to express different architectural languages, which makes a balance between economic and aesthetic and interprets the building space and environment.

■ 建筑设计

建筑以简约现代风格为主，在立面造型设计上注重细节的表达，强调建筑形式、风格、比例的统一塑造。项目以完整的沿街面、大型的广场等大气姿态，来诠释建筑与城市界面的关系。项目设计还注重发挥材质对于建筑的烘托作用，选择适当材质表达不同的建筑语言，兼顾经济与美观，利用现代建筑工艺及材料对建筑空间与环境进行解读。

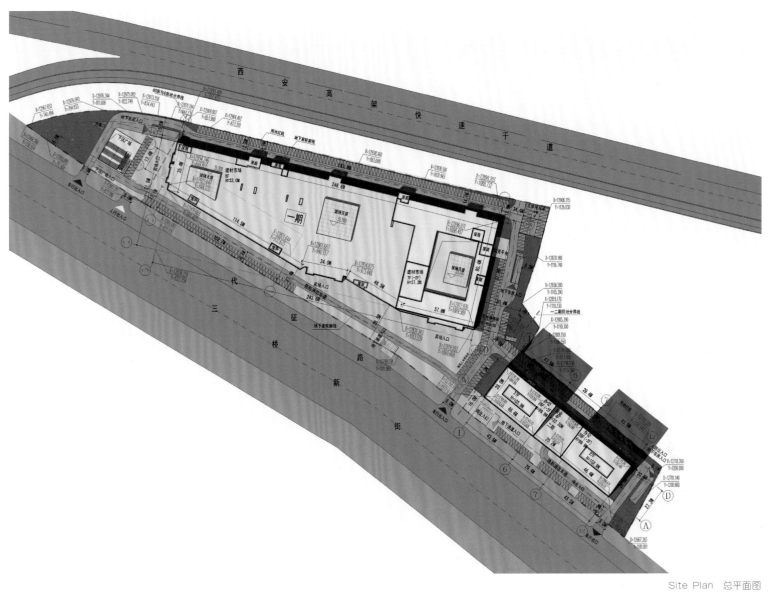

Site Plan 总平面图

067

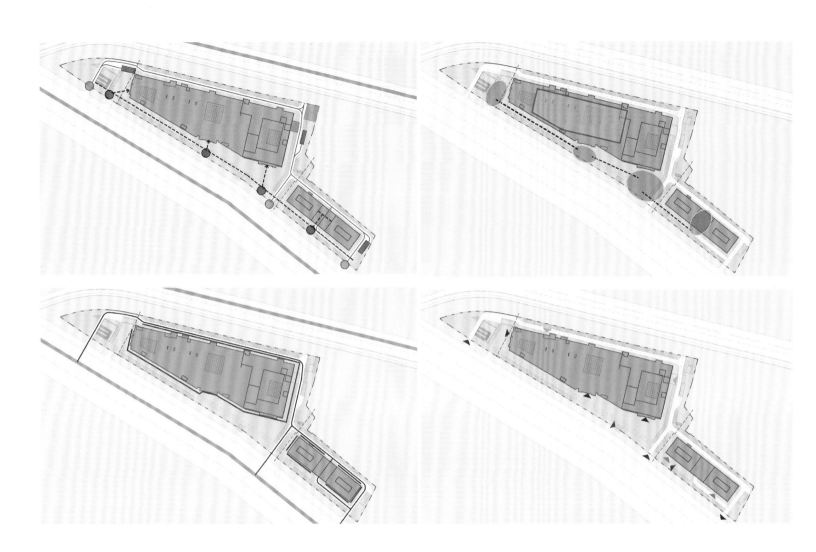

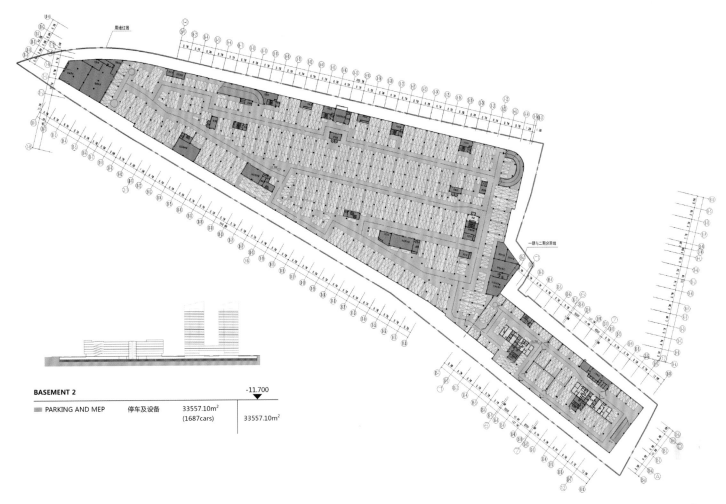

BASEMENT 2

■ PARKING AND MEP 停车及设备 33557.10m² (1687cars)   −11.700   33557.10m²

Basement Two Plan 地下二层平面图

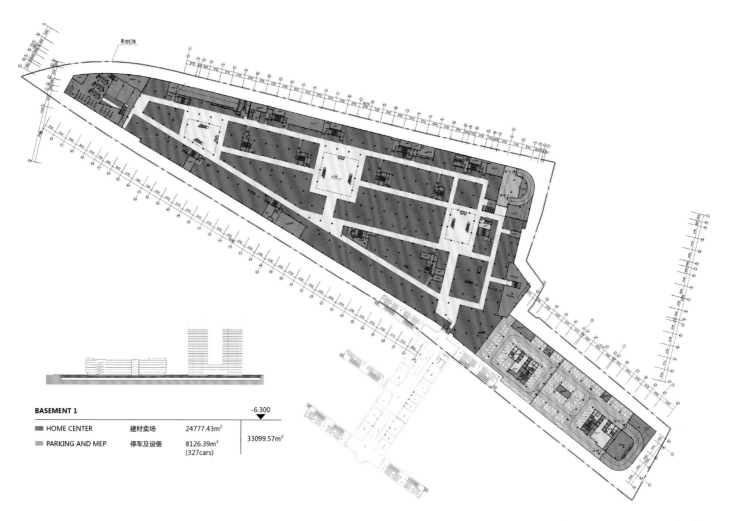

BASEMENT 1 -6.300

■ HOME CENTER 建材卖场 24777.43m²
■ PARKING AND MEP 停车及设备 8126.39m² (327cars)

33099.57m²

Basement Floor Plan 地下层平面图

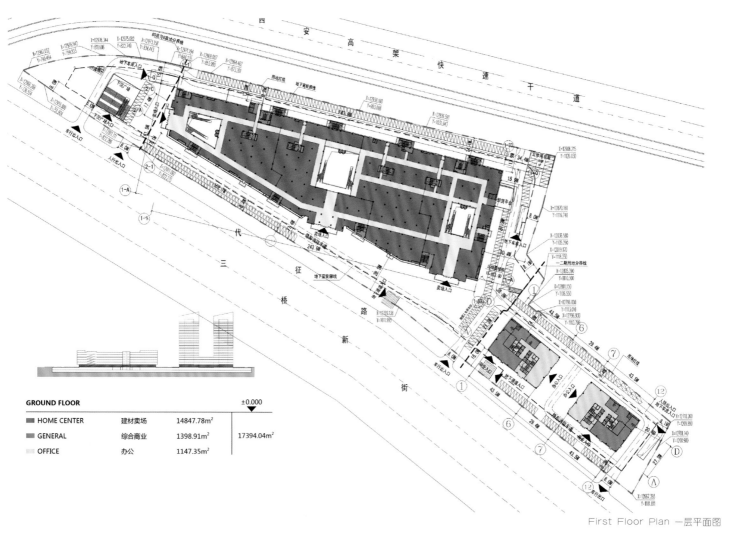

GROUND FLOOR ±0.000

■ HOME CENTER 建材卖场 14847.78m²
■ GENERAL 综合商业 1398.91m²
■ OFFICE 办公 1147.35m²

17394.04m²

First Floor Plan 一层平面图

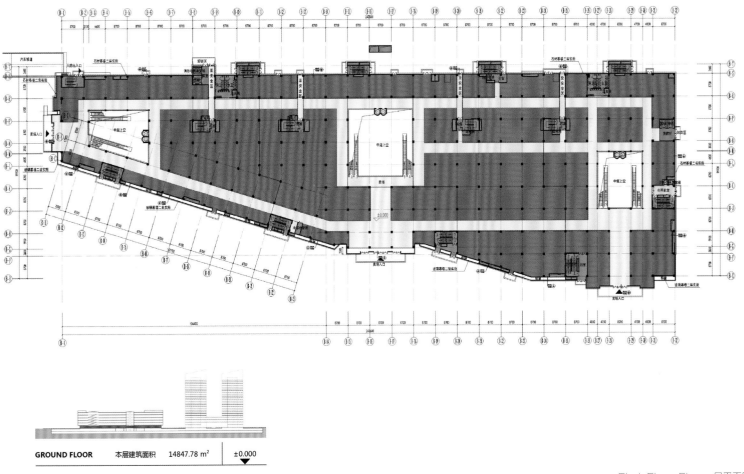

GROUND FLOOR　本层建筑面积　14847.78 m²　±0.000

First Floor Plan 一层平面图

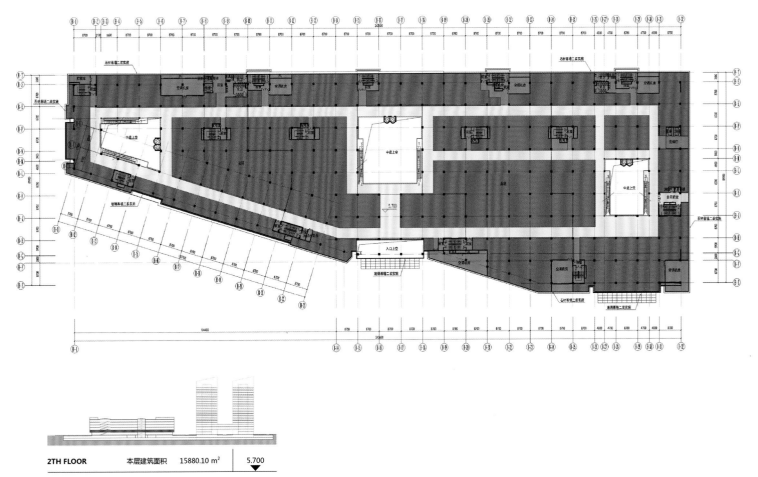

2TH FLOOR　本层建筑面积　15880.10 m²　5.700

Second Floor Plan 二层平面图

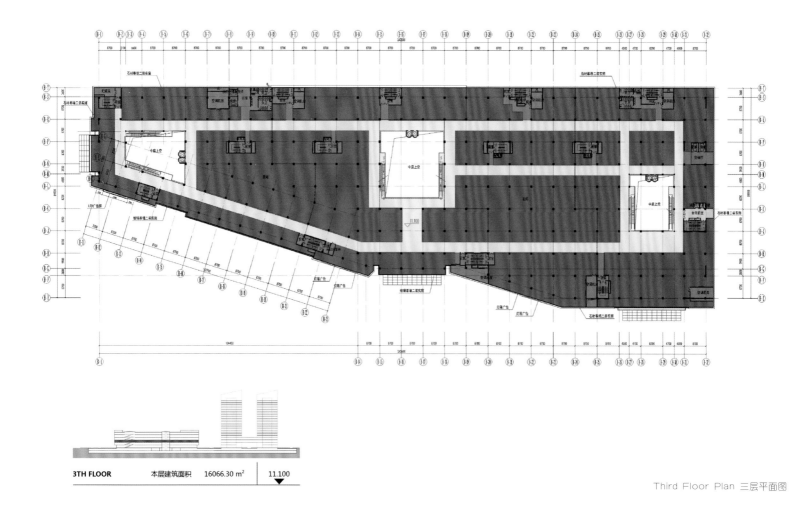

3TH FLOOR　　本层建筑面积　16066.30 m²　　11.100

Third Floor Plan 三层平面图

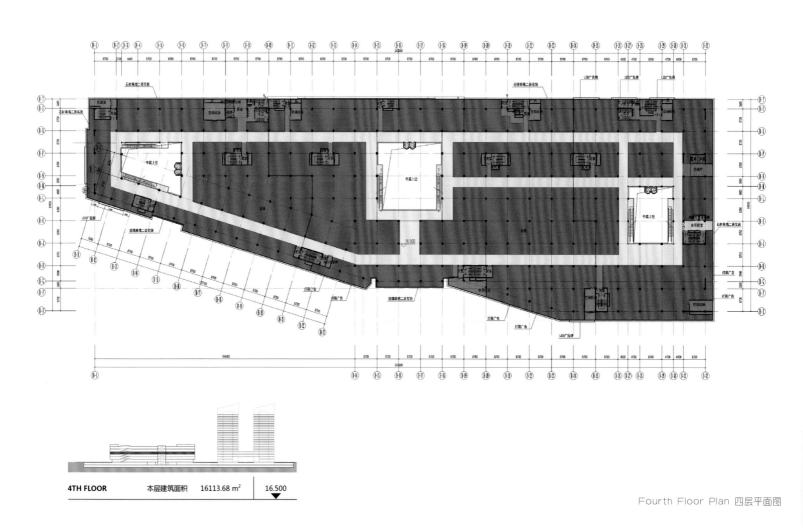

4TH FLOOR　　本层建筑面积　16113.68 m²　　16.500

Fourth Floor Plan 四层平面图

072

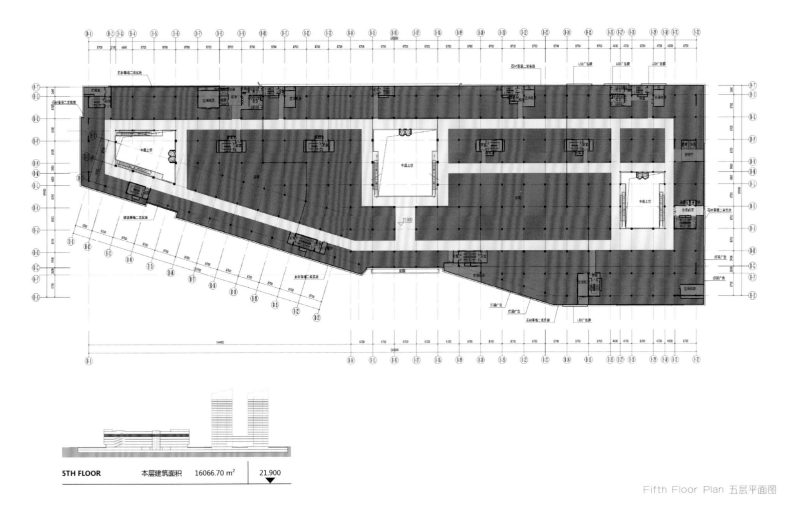

5TH FLOOR 本层建筑面积 16066.70 m² 21.900

Fifth Floor Plan 五层平面图

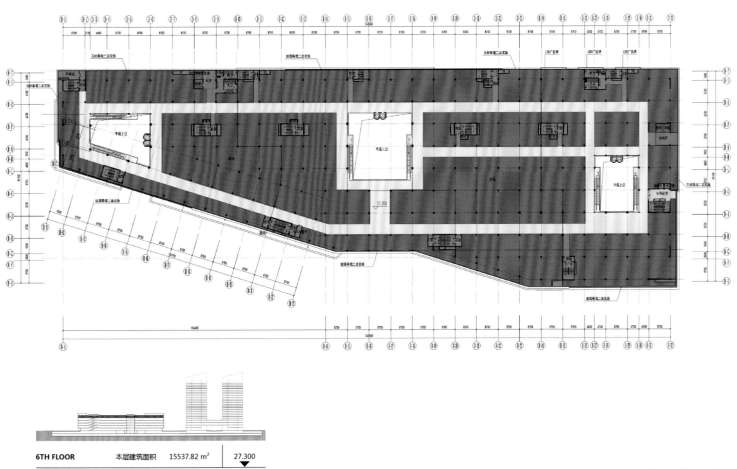

6TH FLOOR 本层建筑面积 15537.82 m² 27.300

Sixth Floor Plan 六层平面图

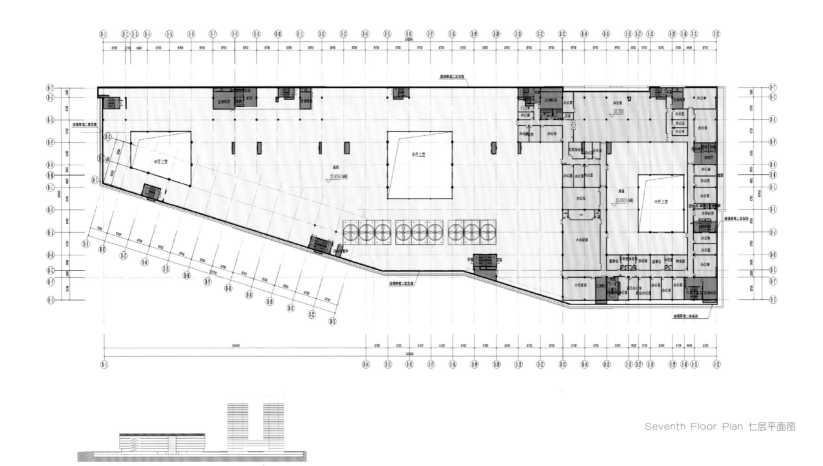

Seventh Floor Plan 七层平面图

7TH FLOOR　　本层建筑面积　4540.89 m²　　▽ 32.700

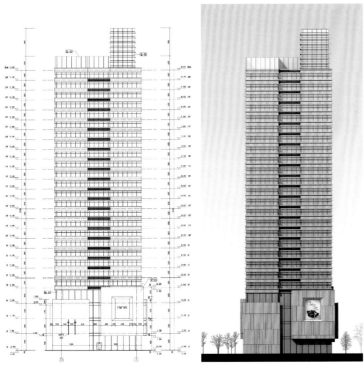

Side Elevation 轴立面图

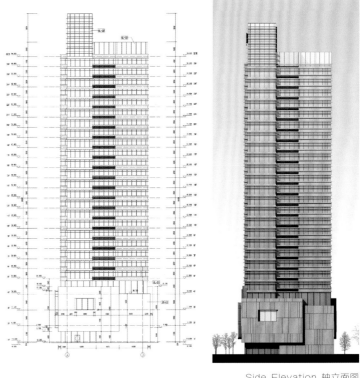

Side Elevation 轴立面图

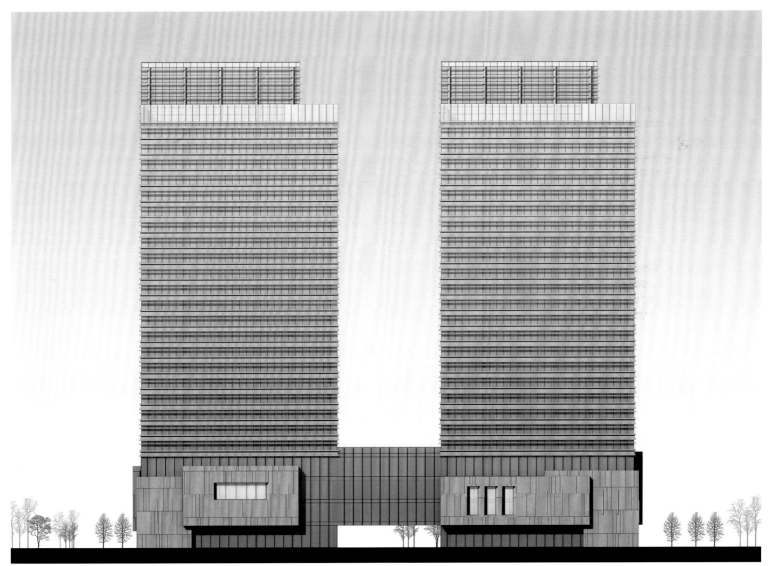

Side Elevation 轴立面图

Side Elevation 轴立面图

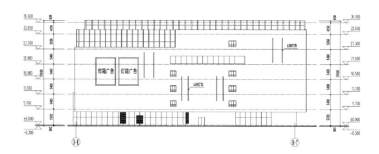
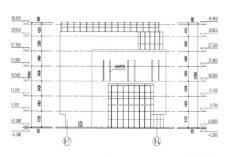

Side Elevation 轴立面图

Side Elevation 轴立面图

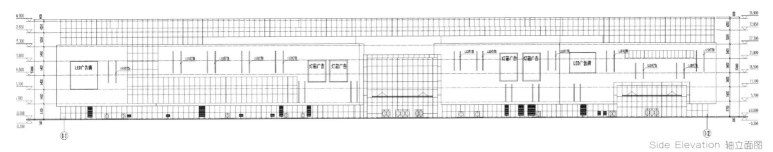

Side Elevation 轴立面图

Side Elevation 轴立面图

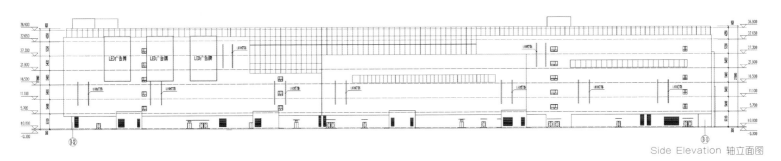

Side Elevation 轴立面图

Side Elevation 轴立面图

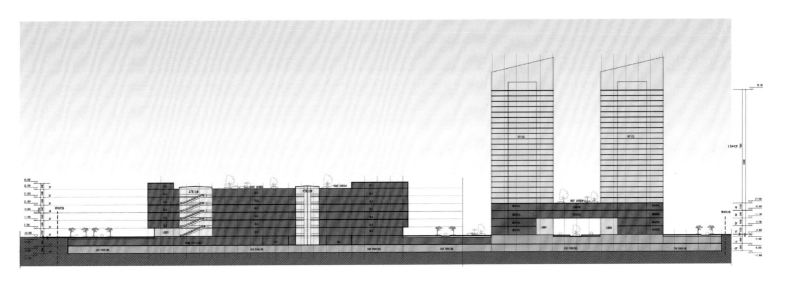

077

Shanghai Fengxian Nanqiao Powerlong City Square

上海奉贤南桥宝龙城市广场

Keywords 关键词

Roof Garden
屋顶花园

Intersection
动线交汇

Dynamic Square
活力广场

Features 项目亮点

The sunken plaza enclosed by the shopping mall and the outdoor pedestrian street plays the core role in the plan, where forms the intersection of the indoor commercial moving lines and the commercial street's outdoor moving lines. The performance stage is set inside the sunken plaza where the wonderful commercial activities can be staged.

由购物中心及室外步行街围合出的下沉广场是规划中的核心部位，在这里，购物中心的室内商业动线和商业街的室外动线形成了交汇。在下沉广场中设置有观演的舞台，精彩丰富的商业活动可以在这里上演。

Location: Fengxian, Shanghai, China
Architectural Design: CCI Architecture Design & Consulting Co., Ltd.
Land Area: 40,687.5 m²
Floor Area: 180,245 m²

项目地点：中国上海市奉贤区
设计单位：上海新外建工程设计与顾问有限公司
占地面积：40 687.5 m²
建筑面积：180 245 m²

■ Overview

Shanghai Fengxian Nanqiao Powerlong City Square is located in the southwest corner of the intersection of Tuannan Road and East Ring Road in Shanghai Fengxian Nanqiao Town. The project was designed to vision a high-end integrated commercial project. Once completed, the square will create a brand-new "living & entertainment" establishment for the residents around.

■ 项目概况

项目位于上海奉贤南桥镇团南公路与环城东路交叉口西南角。定位为高端综合性商业项目。项目建成后将为周边居民创造一个全新的"生活娱乐"场所。

■ Function Layout

The "entertainment establishment" in the project is composed of the shopping center, the outdoor commercial street and underground commercial street. The shopping mall contains a variety of formats, such as living boutiques, anchor stores, food and beverage, entertainment and cinemas with its ground floor full open as a display surface along the street which sets three main entrances and exits. The ground floor of the outdoor commercial street's main format are small shops while the second floor with separate entrance is mainly consists of larger shops. The "living establishments" in the project consists of four office towers located above the shopping mall and the commercial street, and on its towers we could fully enjoy the podium roof garden and the nearby commercial landscape.

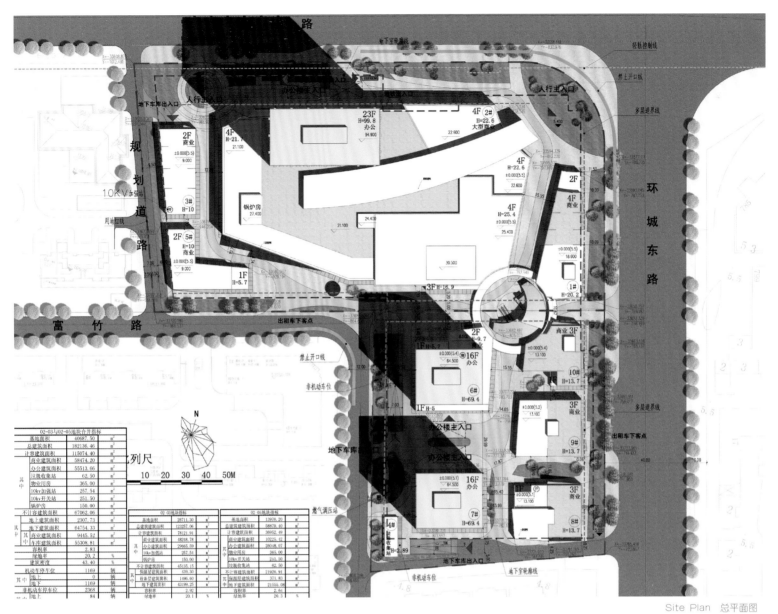

Site Plan 总平面图

图例 Legend	
	日照1小时 SUNLIGHT 1 HOUR
	日照2小时 SUNLIGHT 2 HOUR
	日照3小时 SUNLIGHT 3 HOUR
	日照4小时 SUNLIGHT 4 HOUR
	日照5小时 SUNLIGHT 5 HOUR
	日照5小时 SUNLIGHT 5 HOUR

Daylight Analysis 日照分析图

Elevation of Building 2# 二号楼立面图

Elevation of Building 2# 二号楼立面图

Side Elevation 轴立面图

Side Elevation 轴立面图

■ 功能布局

项目中的"娱乐建设"由购物中心、室外商业街和地下商业街构成。购物中心包含生活名品店、主力店、餐饮、娱乐以及电影院等多种业态，其底层为全开放的沿街展示面，并设置有三个主要的出入口。室外商业街底层以小商铺为主要业态，二层以上为较大的店铺，有独立的出入口。项目中的"生活建设"由四栋办公塔楼组成，坐落在购物中心和商业街的上方，在塔楼上人们可以充分享受裙房的屋顶花园以及邻近的商业景观。

■ Core Role & Landscape Design

The sunken plaza enclosed by the shopping mall and the outdoor pedestrian street plays the core role in the plan, where forms the intersection of the indoor commercial moving lines and the commercial street's outdoor moving lines. The performance stage is set inside the sunken plaza where the wonderful commercial activities can be staged. Connected with the underground business street, the fan-shaped big steps are able to reach the underground garage directly. Because of its "core" role, all the commercial flow of people could reach here through different levels of the moving lines. In terms of landscape design, water features, lighting and music create a vibrant plaza space. A modern lightweight glass canopy is designed at the top of the big step, whose image is similar to the traditional fishing boats.

■ 核心设计与景观设计

由购物中心及室外步行街围合出的下沉广场是规划中的核心部位，在这里，购物中心的室内商业动线和商业街的室外动线形成了交汇。在下沉广场中设置有观演舞台，精彩丰富的商业活动可以在这里上演。扇形的大台阶能够与地下商业街相连通，同时也能够直接到达地下车库。下沉广场位于项目的"核心"位置，因此所有的商业人流都能够通过不同层面的动线到达此处。在景观设计方面，通过水景、灯光与音乐创造出一个具有活力的广场空间。在大台阶的顶部，设计了现代的轻质玻璃雨棚，其形象与传统的渔船类似。

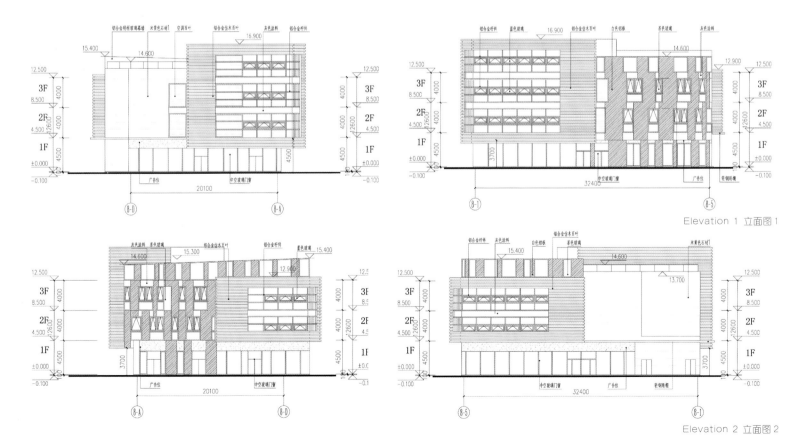

Elevation 1 立面图 1

Elevation 2 立面图 2

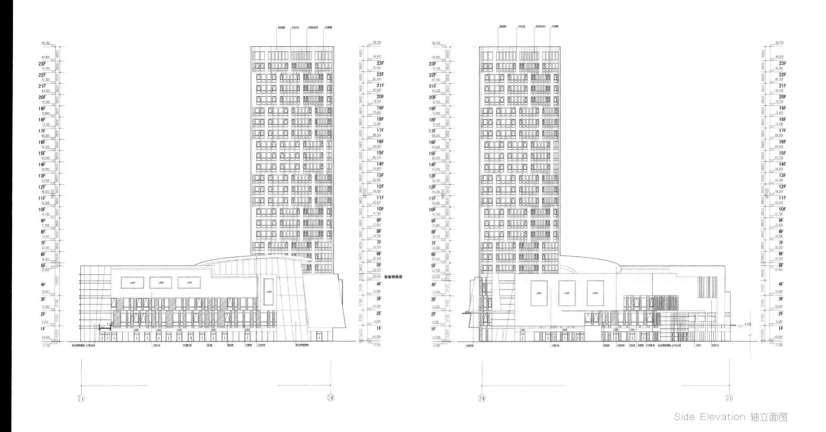

Side Elevation 轴立面图

085

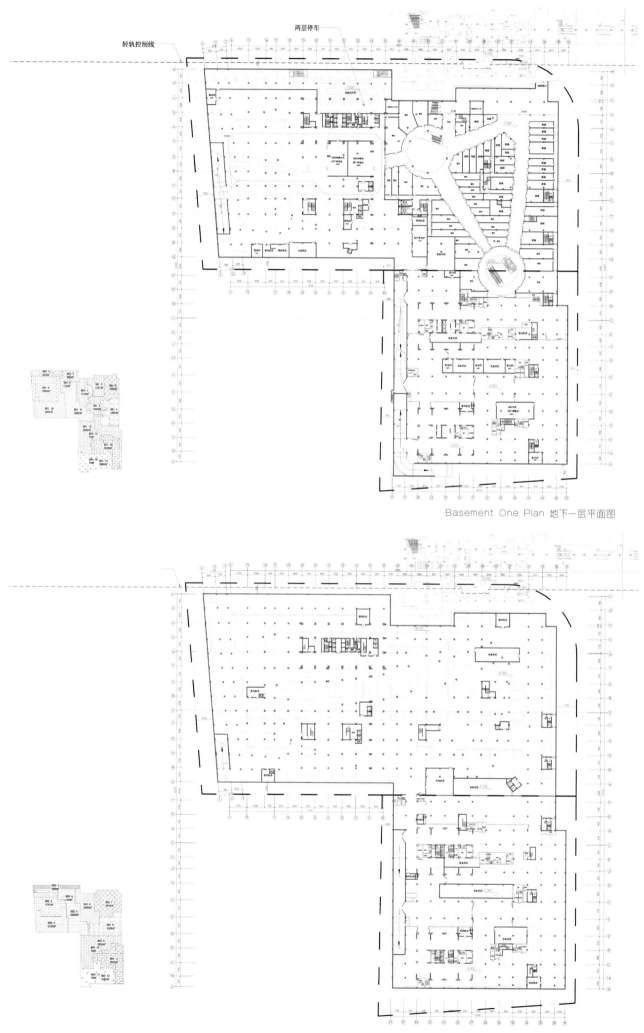

Basement One Plan 地下一层平面图

Basement Two Plan 地下二层平面图

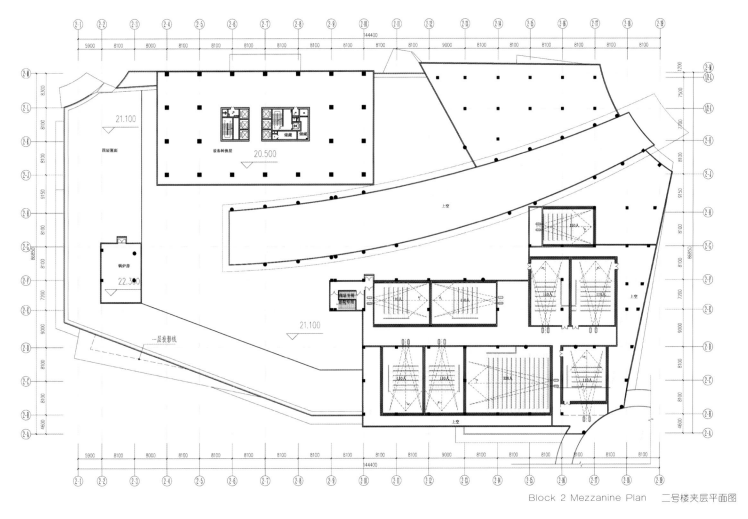

Block 2 Mezzanine Plan 二号楼夹层平面图

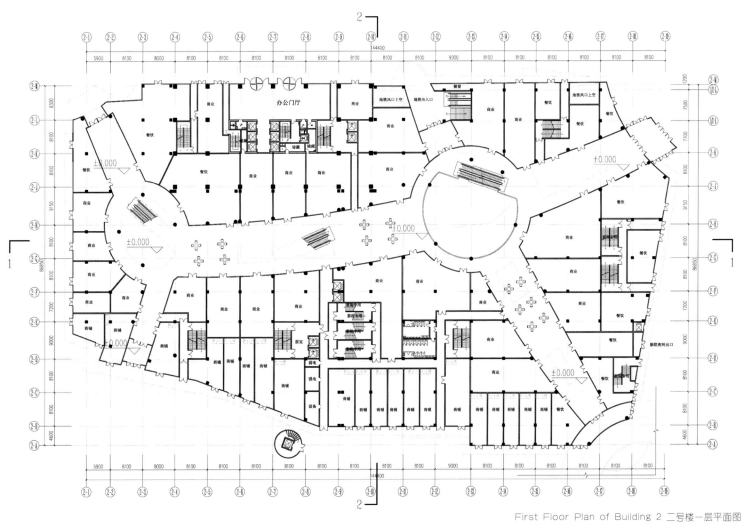

First Floor Plan of Building 2 二号楼一层平面图

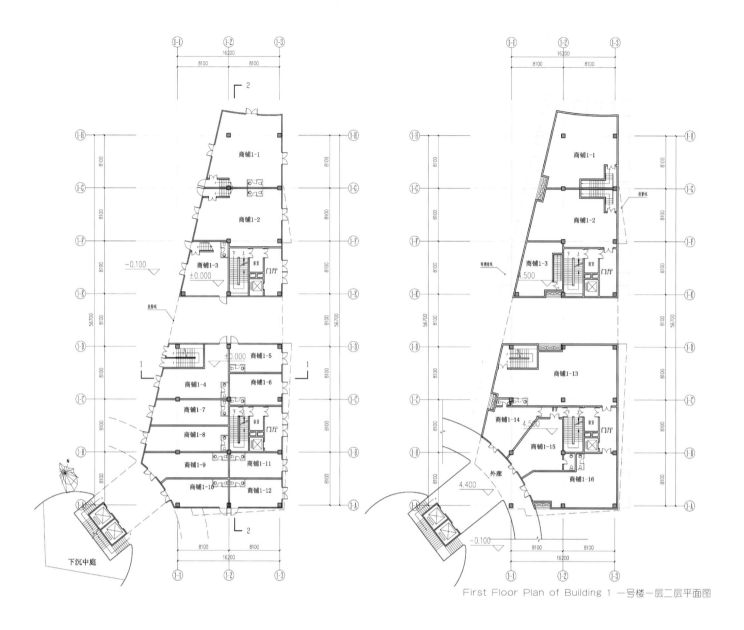

First Floor Plan of Building 1 一号楼一层二层平面图

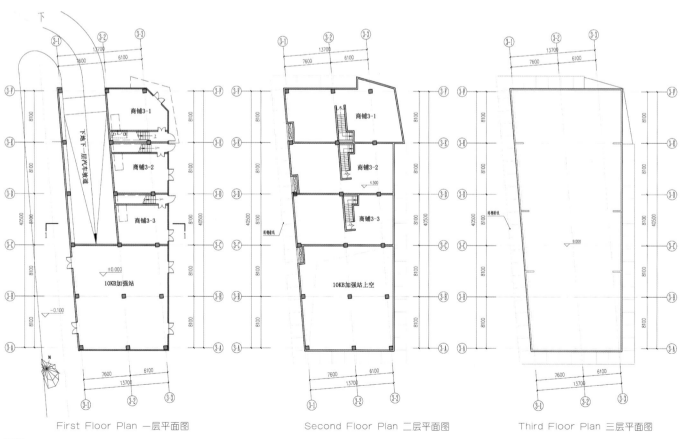

First Floor Plan 一层平面图　　Second Floor Plan 二层平面图　　Third Floor Plan 三层平面图

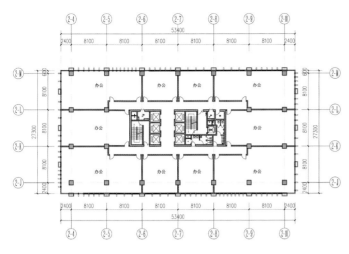

11th, 13th, 17th and 19th Floor Plan 11、13、17、19层平面图

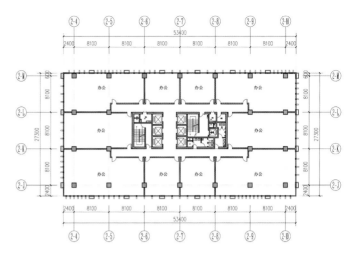

14th and 16th Floor Plan 14、16层平面图

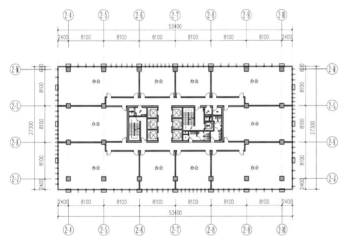

5th, 7th, 23th and 24th Floor Plan 5、7、23、24层平面图

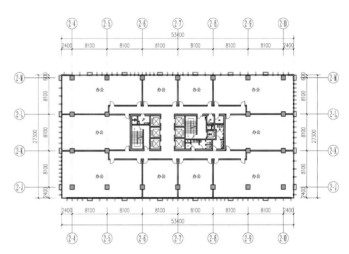

8th, 10th, 14th and 16th Floor Plan 8、10、21、22层平面图

Kunming Nanshi Central Golden Estate Project – Phase II

昆明南市中央金座二期

Keywords 关键词

Urban Characteristics
城市特征

Rich Facade
立面丰富

Beautifully Shaped
造型优美

Location: Kunming, Yunnan, China
Architectural Design: CCI Architecture Design & Consulting Co., Ltd.
Land Area: 24,296.95 m²
Floor Area: About 29,700 m²

项目地点：中国云南省昆明市
设计单位：上海新外建工程设计与顾问有限公司
占地面积：24 296.95 m²
建筑面积：约29 700 m²

Features 项目亮点

The project includes a shopping center, a 5A office building, a spa, and an underground business area connected to the subway station, and the various parts are interdependent and bring out the best of each other. The design integrates the city's characteristics, tries to find the best meeting point between the places and people, and showcases these features through the project nodes.

项目包含购物中心、5A写字楼、温泉浴场以及与地铁站相连的地下商业，各部分之间相互依存、相得益彰。设计结合城市特色，力图寻找场所与人之间的最优契合点，并通过项目节点展现这些特征。

■ **Overview**

The Kunming Nanshi Central Golden Estate Project – Phase II is located at the junction of Dianchi Road and Guangfu Road, about 2.4 km from the Second Ring Road of Kunming. The convenient transportation accessible to all directions becomes an important traffic advantage to build a commercial complex. Furthermore, the surroundings are mostly residential areas with abundant living facilities, which is a favorable condition to commercial projects.

■ **项目概况**

昆明南市中央金座二期项目位于滇池路与广福路交接处，距离昆明市二环路约2.4 km，交通便利，是打造商业综合体的重要交通优势；周边大多为住宅小区，生活配套齐全，成为做商业项目的有利条件。

■ **Project Orientation**

The project is a large mixed use complex with a collection of retails, offices, restaurants, entertainment facilities and other commercial formats. The project includes a shopping center, a 5A office building, a spa, and an underground business area connected to the subway station, and the various parts are interdependent and bring out the best of each other. As a continuation of the first phase, the second phase project enhances the commercial operation of the first one, and strengthens the Golden Estate brand as well.

■ **项目定位**

项目是一座汇集了商业、办公、餐饮、文娱等多业态的大型城市综合体。项目包含购物中心、5A写字楼、温泉浴场以及与地铁站相连的地下商业，各部分之间相互依存、相得益彰。昆明南市中央金座二期作为一期项目的延续，提升了一期项目的商业运营，同时也强化了金座的品牌效应。

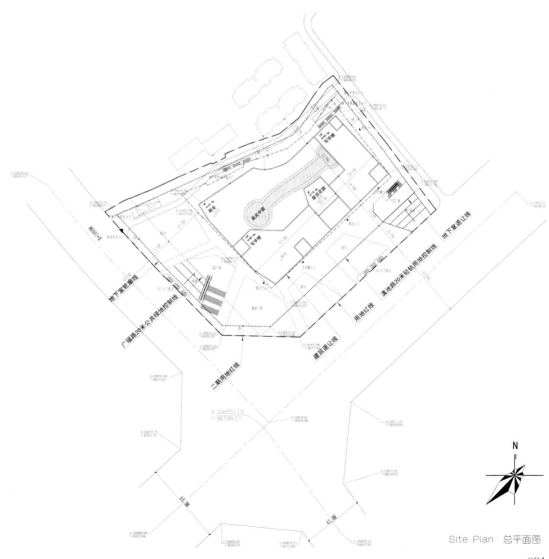

Site Plan 总平面图

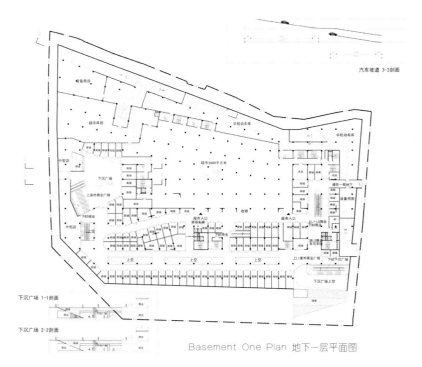
Basement One Plan 地下一层平面图

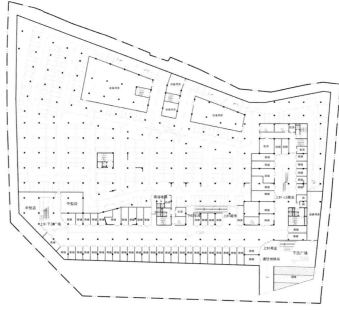
Basement Two Plan 地下二层平面图

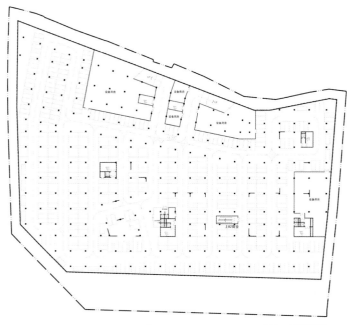
Basement Three Plan 地下三层平面图

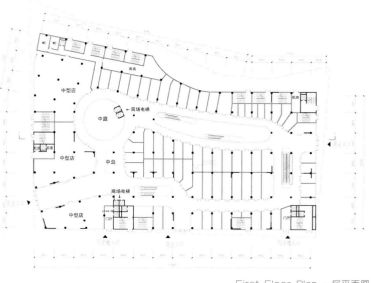
First Floor Plan 一层平面图

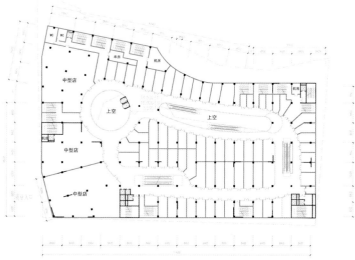
2-4 Floor Plan 2-4层平面图

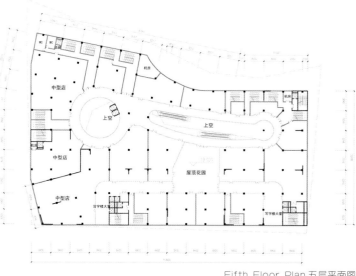
Fifth Floor Plan 五层平面图

Design Characteristics

The design integrates the city's characteristics, tries to find the best meeting point between the places and people, and showcases these features through the project nodes.

Characteristics of the city: four seasons like spring, outdoor living, adventurous spirits, leisure, close to nature, diversity, regionality.

Green roof platform: providing outdoor space, increasing connection to nature, reducing the effect of Heat Island, promoting outdoor retail environment.

Sunken plaza: a multi-level ground floor entrance, accessibility and convenience of commercial space, shaping different urban interfaces, spatial experience.

Spa: slow life, rich format levels, use of natural resources.

■ 设计特征

设计结合城市特色，力图寻找场所与人之间的最优契合点，并通过项目节点展现这些特征。

城市特征：四季如春、户外生活、冒险精神、悠闲、接近自然、多样性、区域性。

屋顶绿化平台：提供室外空间、增加对自然的联系、降低热岛效应、推动户外零售环境。

下沉广场：多层次地面入口、商业空间的可达性和便捷性、塑造不同城市界面、空间体验。

温泉浴场：慢生活、丰富业态层次、对自然资源的利用。

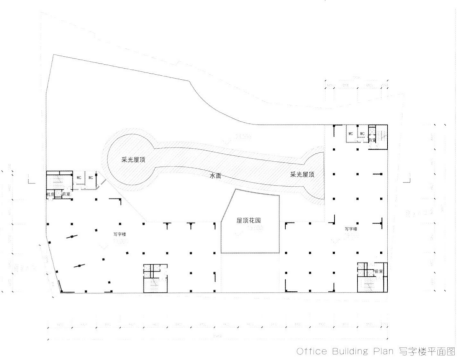

Office Building Plan 写字楼平面图

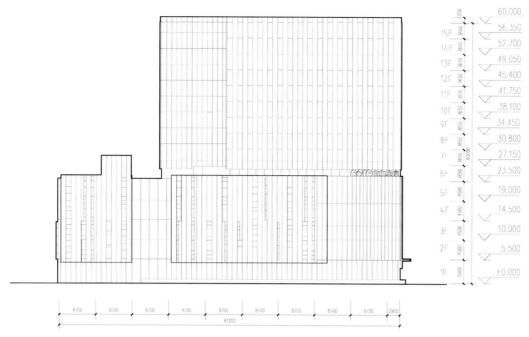

West Elevation 西立面图

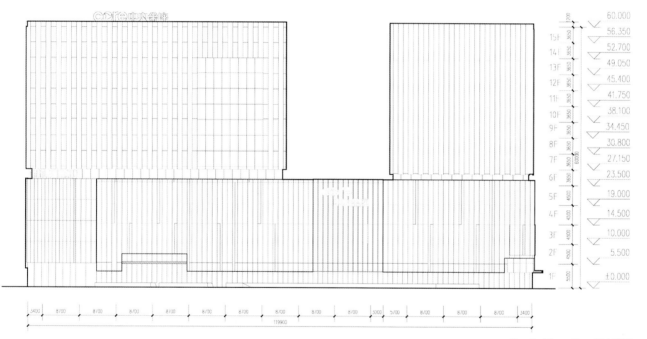

South Elevation 南立面图

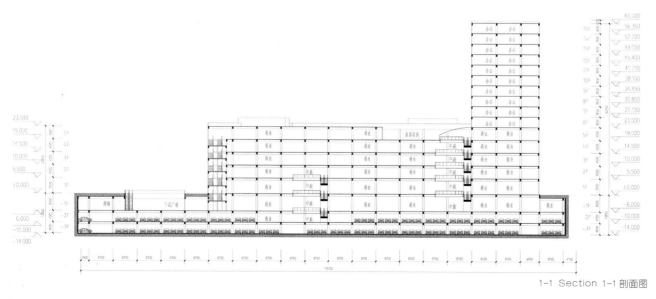

1-1 Section 1-1 剖面图

Qingdao Zhengjian Golden Town
青岛政建金地世纪城

Keywords 关键词

Festive Walk
节庆大道

Commercial Aesthetics
商业美学

Narrative Technique
叙事性手法

Location: Qingdao, Shandong, China.
Architectural Design: SHANGHAI DE-SIGN ARCHITECTURAL DESIGN CO., LTD.
Total Floor Area: 199,825.7 m²
Ground Floor Area: 134,801.7 m²
Underground Floor Area: 65,024 m²

项目地点：中国山东省青岛市
设计公司：鼎世设计集团
总建筑面积：199 825.7 m²
地上建筑面积：134 801.7 m²
地下建筑面积：65 024 m²

Features 项目亮点

In initial planning, designers uphold the concept of "move people first, make them stay then", and they incorporate diversified commercial space with urban public open space in order to support various shopping, catering and recreational activities in the future.

项目在规划初期秉持"先让人感动，再让人停留"的理念，将多样化的商业空间与城市公共开放空间融为一体，以期在未来支持多样丰富的购物、餐饮及娱乐活动。

■ Overview

Located in Qingdao with a limit aviation height of 53 m, the project base has a planned subway line. It accommodates five-star hotel, supermarket, apartment, cinema, catering and children's world which complement with the existing commercial activities to build a commercial center in Chengyang. In planning, the development concept of "commercial aesthetics" is integrated with urban development to create an iconic, high value-added and multi-functional urban commercial product for the commercial core district.

■ 项目概况

项目位于青岛市城阳区，航空限高53 m，基地内有规划地铁线通过。项目在业态上引入五星级酒店、超市、公寓、影院、餐饮娱乐、儿童乐园等，与现有的商业业态形成互补，共同打造成为城阳商贸中心。在规划中将"商业美学"的开发理念与城市发展规划相融合，全力打造该商业核心区的具有城市地标性，高附加值，多功能化的城市商业精品。

■ Business Planning & Spatial Layout

In business planning, the project integrates urban living consumption with characteristic commercial orientation. In business matching, proper and professional proportions of hotel, retail, catering and recreation are guaranteed to ensure the successful operation of commercial project.

In spatial layout, by learning the design concept of the Singapore "Festive Walk", a situational three-dimensional ecological shopping street links up different theme nodes throughout the entire land organically. Through commercial pattern to attract bustling people flow to interpret a never-ending carnival.

■ 商业规划与空间布局

在商业规划层面，项目立足城市区域生活消费，融入特色商业定位。在商业配比上保证酒店、商业、餐饮、娱乐等多功能业态的合理专业比例，保证商业项目的成功运营。

在空间布局上引入了新加坡"节庆大道"的设计理念，用一条集购物、休闲、景观为一体的情景化立体生态购物街将整个地块不同主题的节点有机地串联起来，通过商业格局的打造，导入熙熙人流，以瑰丽喧嚣烘托出一场永远也不会落幕的嘉年华。

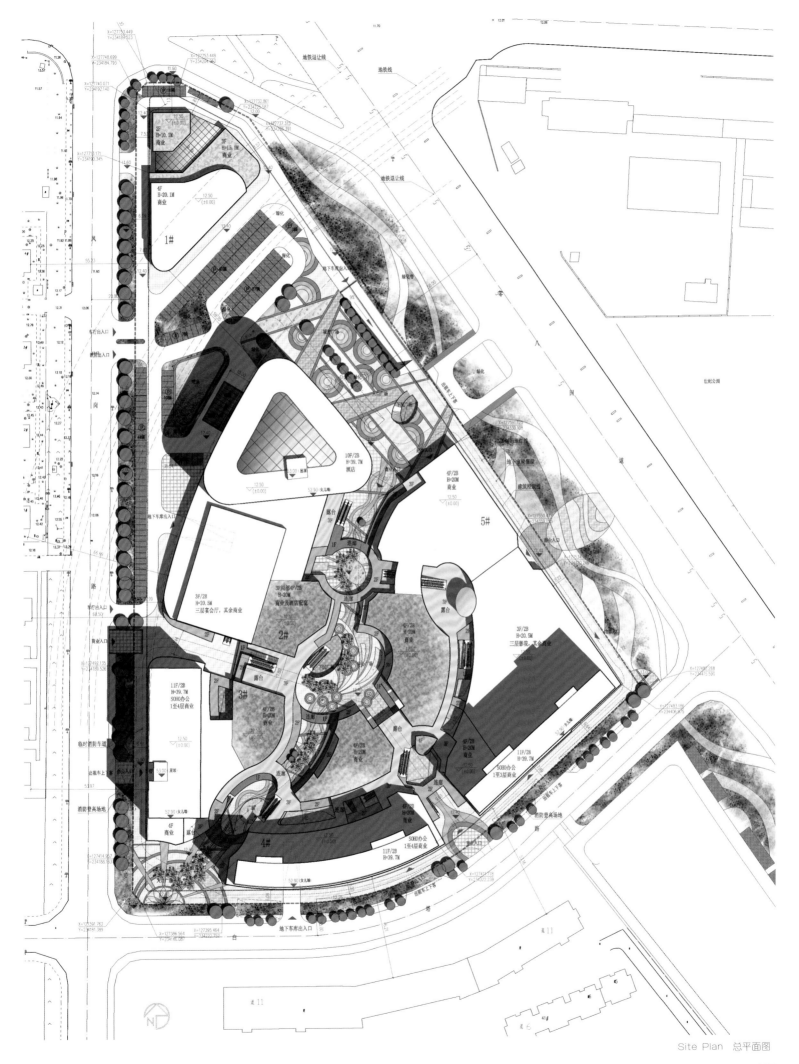

Site Plan 总平面图

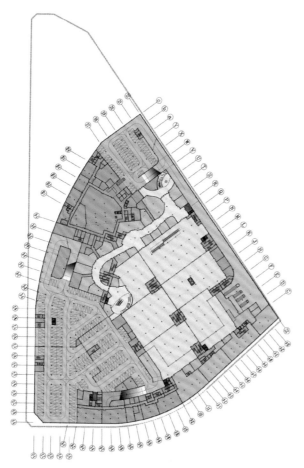

Floor Plan 1 平面图 1

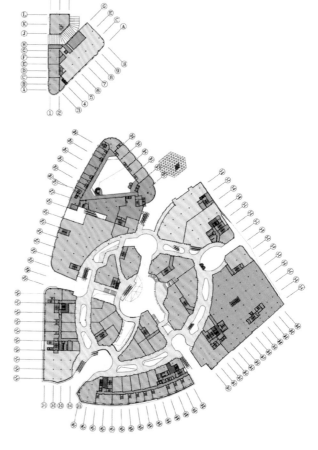

Floor Plan 2 平面图 2

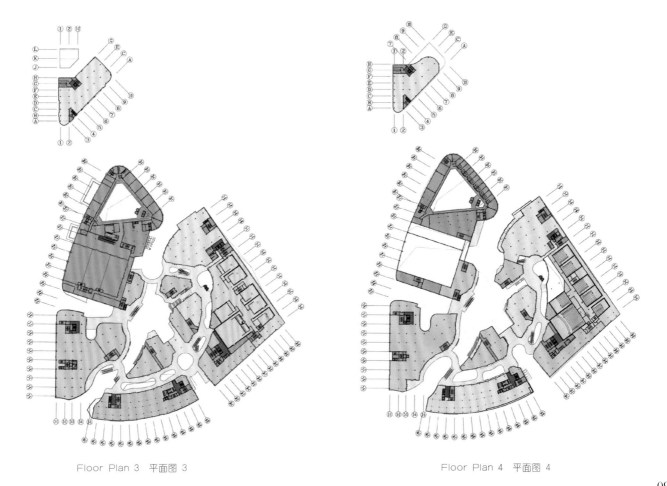

Floor Plan 3 平面图 3

Floor Plan 4 平面图 4

■ Architectural Design

In initial planning, designers uphold the concept of "move people first, make them stay then", and they incorporate diversified commercial space with urban public open space in order to support various shopping, catering and recreational activities in the future. The planning space concept takes "Festival Walk" as the background and spreads "Celebration Square" and "Holiday Square" along it. The designs of facades and scenes are presented with narrative technique to jointly build distinctive features of the entire project in the premise of a happy and peaceful atmosphere.

■ 建筑设计

项目在规划初期秉持"先让人感动，再让人停留"的理念，将多样化的商业空间与城市公共开放空间融为一体，以期在未来支持多样丰富的购物餐饮娱乐活动。规划空间理念都以"节庆大道"为背景，从"庆典广场"到"假日广场"，立面及场景的设计都用叙事性的精工手法，以营造欢乐祥和的氛围为前提，共同塑造整个项目的鲜明特色。

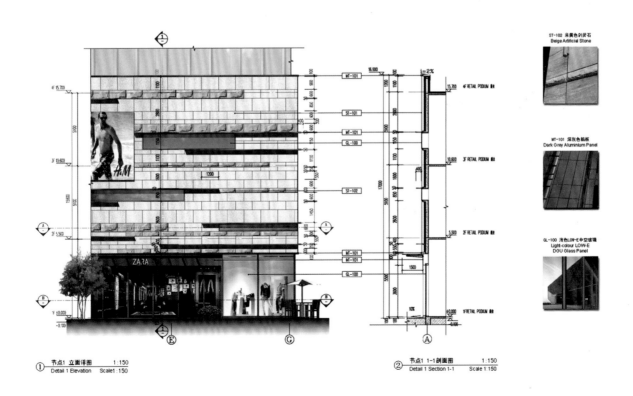

① 节点1 立面详图　　1:150
　　Detail 1 Elevation　Scale 1:150

② 节点1 1-1剖面图　　1:150
　　Detail 1 Section 1-1　Scale 1:150

ST-102 米黄色剑麦石
Beige Artificial Stone

MT-101 深灰色铝板
Dark Grey Aluminium Panel

GL-100 浅色LOW-E中空玻璃
Light-colour LOW-E
DGU Glass Panel

Shanghai Jiading New City Plot A131
上海嘉定新城 A131 地块

Keywords 关键词

Reasonable Layout
布局合理

Tower Design
塔楼设计

People Flow Line
人流动线

Features 项目亮点

In accordance with modern business idea, the project aims to create a modern commercial complex with reasonable layout, convenient traffic, efficient operation and beautiful environment.

项目旨在以现代商业理念贯穿设计始终，进行合理功能布局，打造布局合理、交通便捷、运转高效、环境优美的现代商业综合体。

Location: Jiading, Shanghai, China
Architectural Design: CCI Architecture Design & Consulting Co., Ltd.
Land Area: 18,623.1 m²
Floor Area: 57,592.4 m²

项目地点：中国上海市嘉定区
设计单位：上海新外建工程设计与顾问有限公司
占地面积：18 623.1 m²
建筑面积：57 592.4 m²

■ **Overview**

The project is located in the core area of southern Jiading New city which is one of the three key development areas in Shanghai. As the main plot in the development area and the concentrated area comprised of residential community, culture and commerce, the project boasts a planning land area of 18,623.1 m², an overground floor area of 57,592.4 m² and a plot ratio of 3.0. In accordance with modern business idea, the project aims to create a modern commercial complex with reasonable layout, convenient traffic, efficient operation and beautiful environment.

■ **项目概况**

项目地块位于上海三大重点开发区域之一——嘉定新城主城区的南部城区核心区域，属于整个嘉定新城板块的主要发展区及居住、文化、商业的集中区域。本项目规划用地面积18 623.1 m²，地上建筑面积约为57 592.4 m²，容积率为3.0。项目旨在以现代商业理念贯穿设计始终，进行合理功能布局，打造布局合理、交通便捷、运转高效、环境优美的现代商业综合体。

■ **Function Composition**

The project is divided into two parts, "recreation place" and "living place". "Recreation place" mainly refers to the commercial part which is composed of the outdoor commercial street and underground supermarket with its ground floor full open as a display surface along the street where three main entrances and exits are set. The ground floor of the outdoor commercial street's main format is small shops while the second floor with separate entrance is mainly consists of larger shops. The "living place" in the project refers to the apartment-office part which is composed of two office towers located above the commercial street, and on its towers we could fully enjoy the surrounding greenery and nearby commercial landscape. In addition, the project is provided with underground garage.

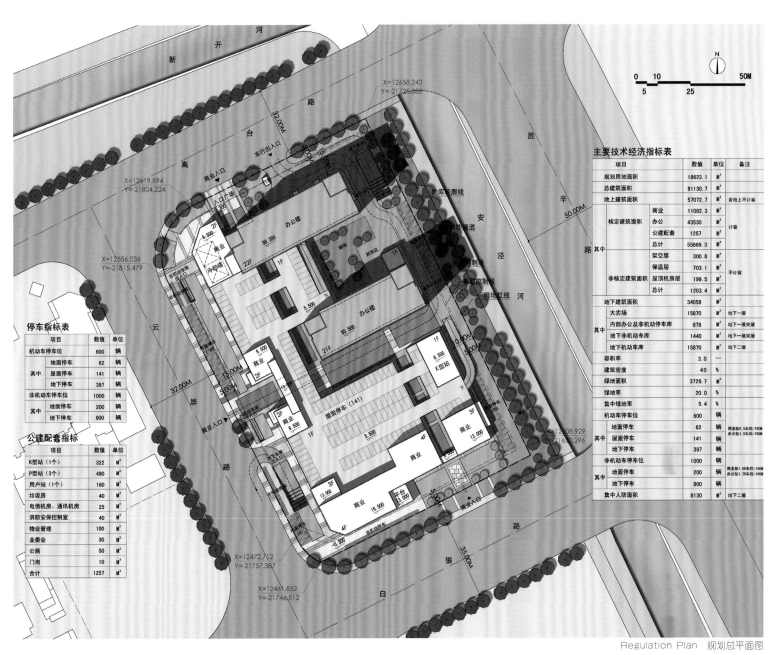

Regulation Plan 规划总平面图

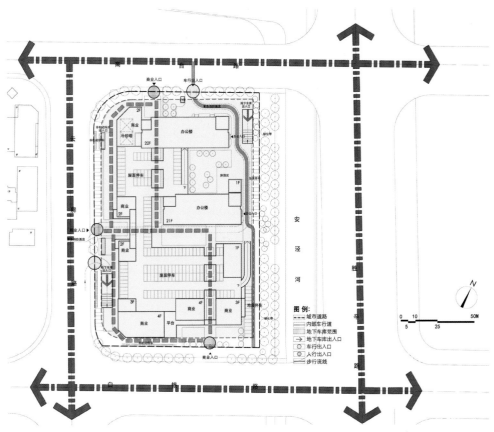

Road Traffic Analysis Diagram 道路交通分析图

■ 功能构成

项目分为"娱乐场所"和"生活场所"两部分。其中"娱乐场所"主要指商业部分，由室外商业街和地下超市构成。其底层为全开放的沿街展示面，并设置三个主要的出入口。室外商业街底层以小商铺为主要业态，二层以上为较大的店铺。而"生活场所"主要指公寓式办公楼部分，由2栋办公塔楼组成，坐落在商业街的上方。在塔楼上人们可以充分享受周围绿化以及邻近的商业景观。项目中设置有地下车库。

■ People Flow Line

Through analyzing the surrounding environment, designers defined the main direction and connection of the people flow and linked up the different function blocks by the dynamic flow which leads people to those functional areas. In the exterior space, dynamic commercial street and visual point are used to guide the people flow, while in the interior space, a series of purposeful consumption pattern is established.

■ 人流动线

在设计中，设计师通过对周边环境的分析确定了人流的主要方向和联系方式，并通过具有动感的人流动线将项目中不同功能的区块紧密地联系起来，将人流按照设计的流线引导到各个功能区。在室外空间通过有活力的商业街以及视觉中心点带引人流的方向，在室内空间则以一系列目的性消费的方式来导引人流动线。

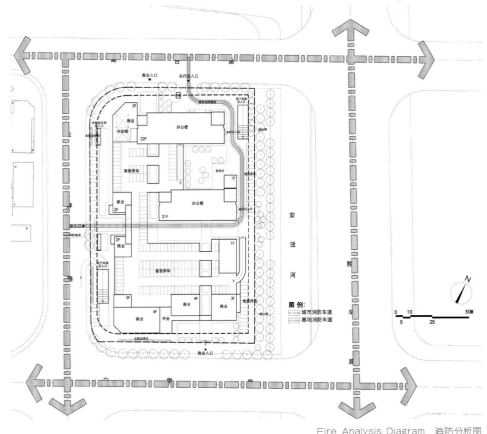

Fire Analysis Diagram 消防分析图

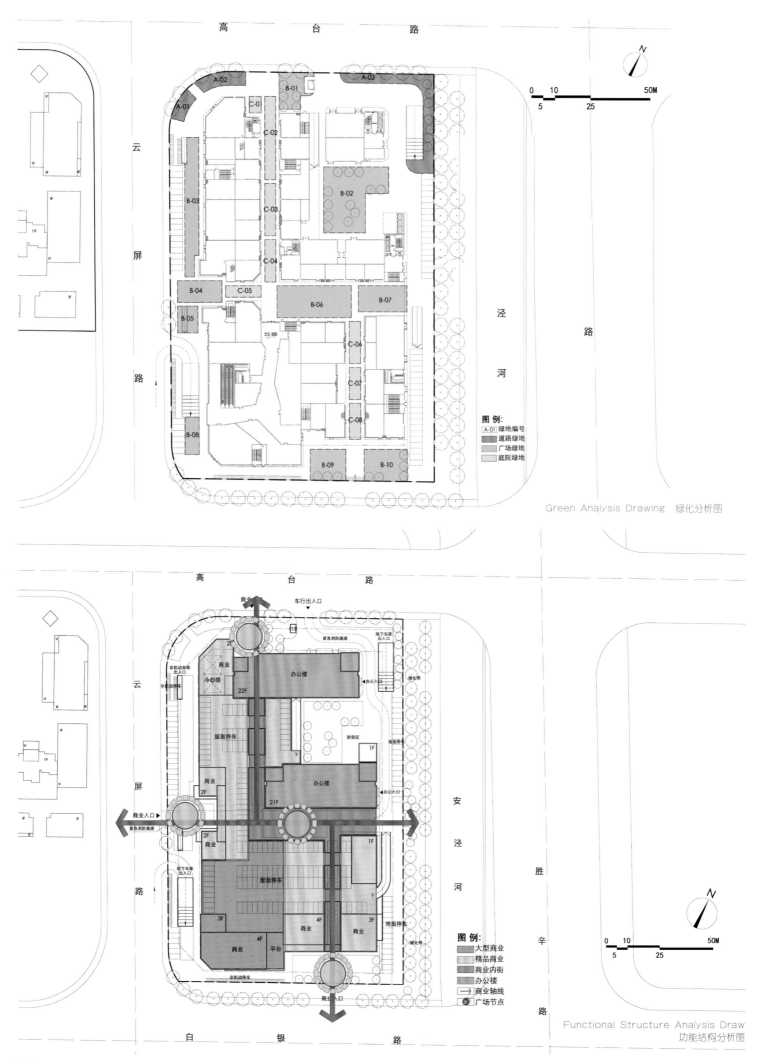

Green Analysis Drawing 绿化分析图

Functional Structure Analysis Draw
功能结构分析图

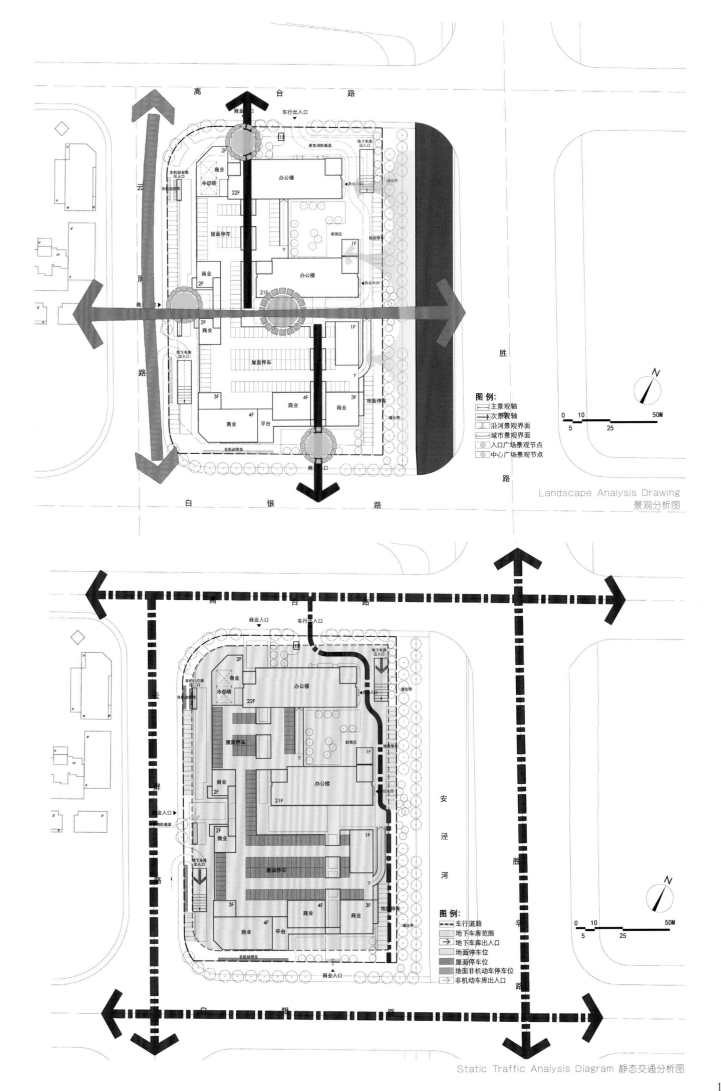

Landscape Analysis Drawing 景观分析图

Static Traffic Analysis Diagram 静态交通分析图

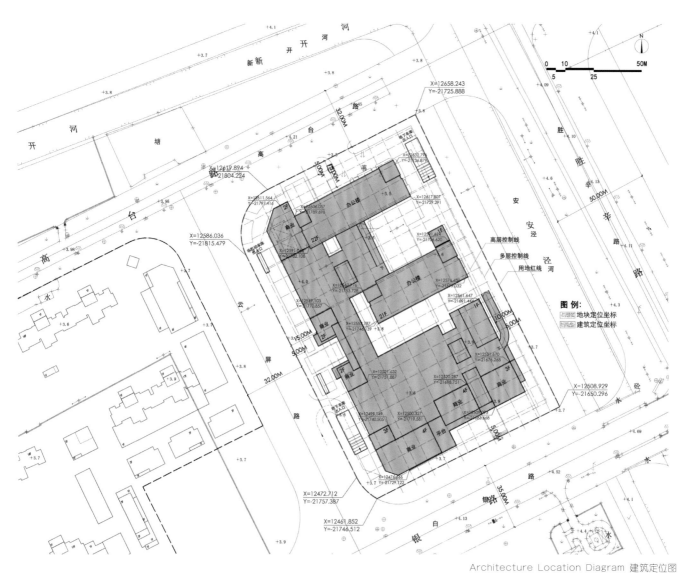

Architecture Location Diagram 建筑定位图

Architecture Interval & Setback Analysis Diagram 建筑间距及退界分析图

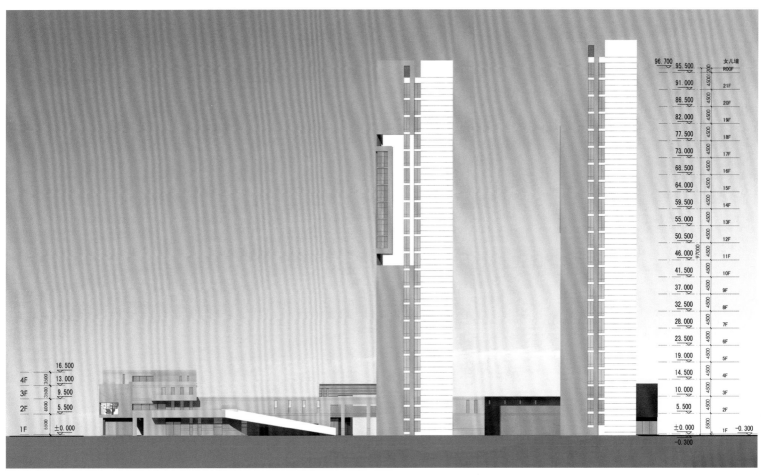

Elevation 1 立面图 1

Elevation 2 立面图 2

Elevation 3 立面图 3

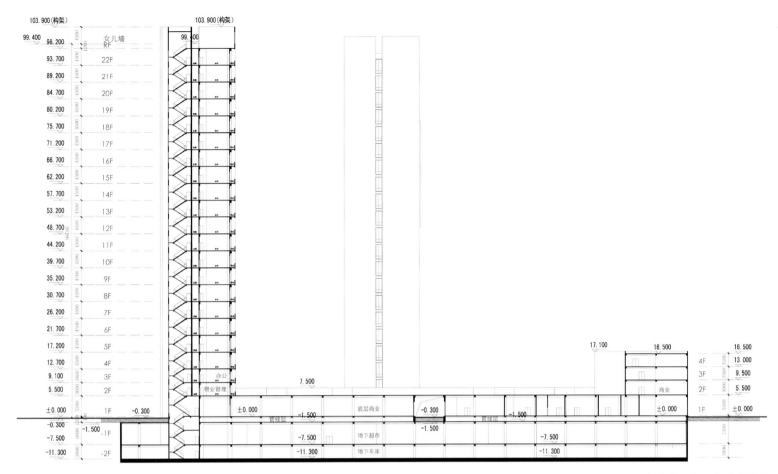

A-A Section A-A 剖面图

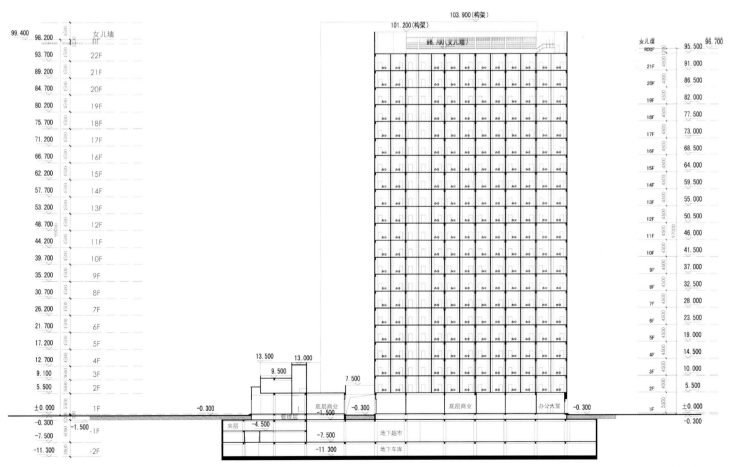

B-B Section B-B 剖面图

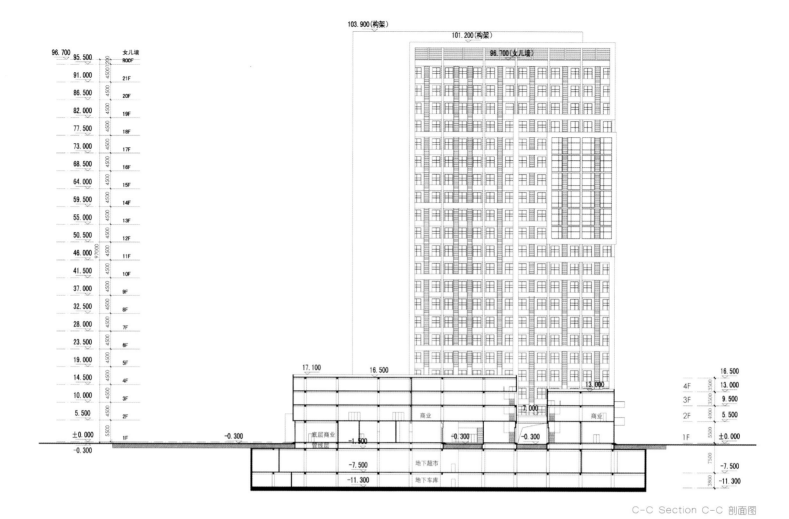

C-C Section C-C 剖面图

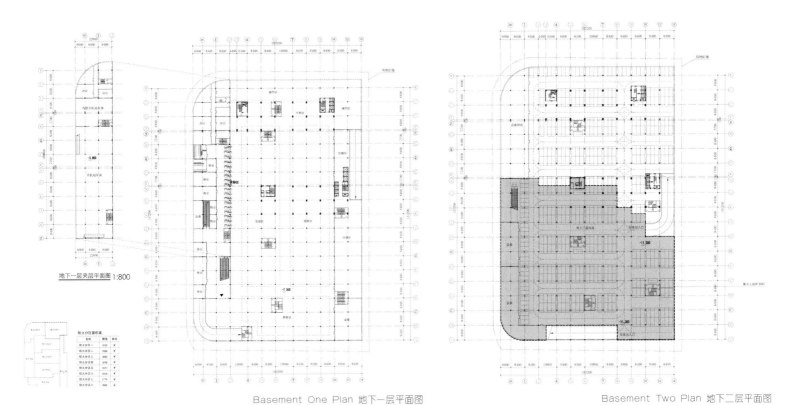

Basement One Plan 地下一层平面图

Basement Two Plan 地下二层平面图

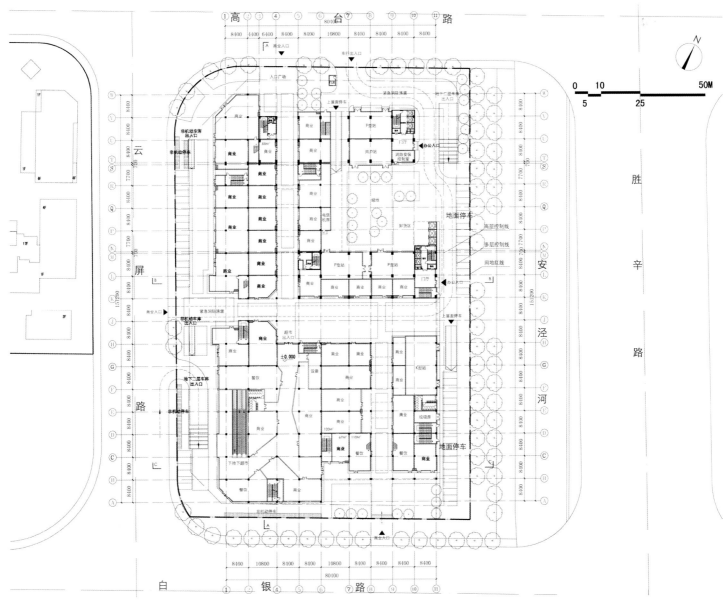

First Floor Plan 一层平面图

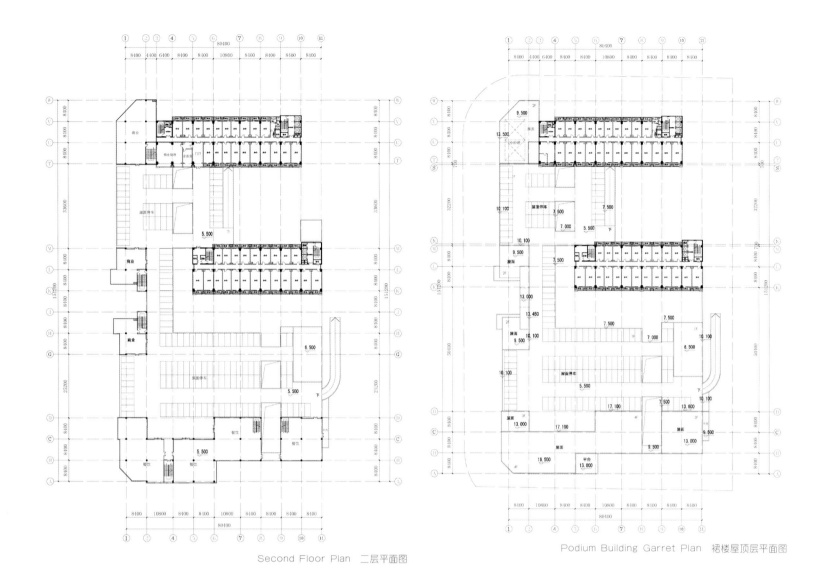

Second Floor Plan 二层平面图

Podium Building Garret Plan 裙楼屋顶平面图

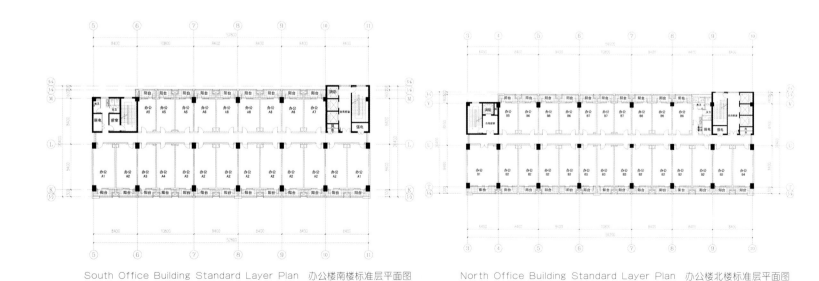

South Office Building Standard Layer Plan 办公楼南楼标准层平面图

North Office Building Standard Layer Plan 办公楼北楼标准层平面图

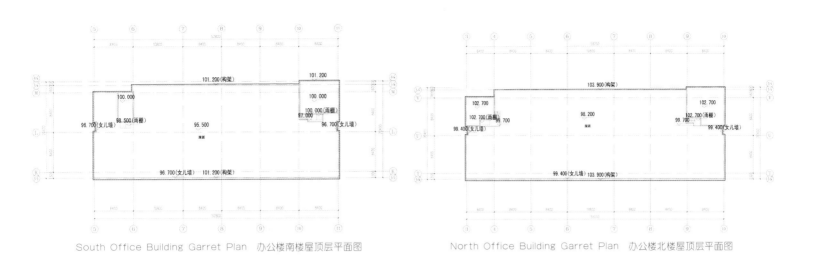

South Office Building Garret Plan　办公楼南楼屋顶层平面图

North Office Building Garret Plan　办公楼北楼屋顶层平面图

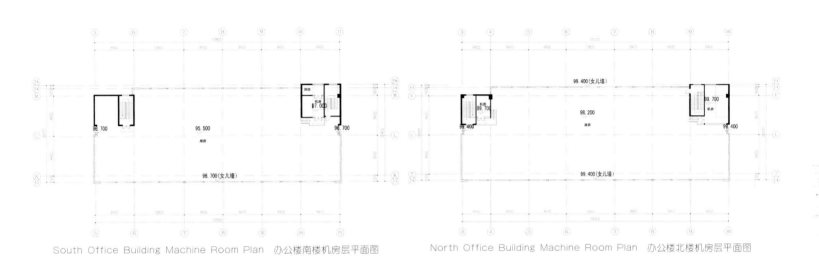

South Office Building Machine Room Plan　办公楼南楼机房层平面图

North Office Building Machine Room Plan　办公楼北楼机房层平面图

Nanhai Xinji Plaza
南海信基广场

Keywords 关键词

Glass Curtain Wall
玻璃幕墙

Sky Garden
中空花园

Backwards Square
退让式广场

Location: Foshan, Guangdong, China
Developer: Xinji Group
Architectural Design Guangzhou Jingsen Engineering Design Consultants Limited
Total (Planning) Site Area: 29,320 m²
Total Floor Area: 187,709 m²
Plot Ratio: 4.9
Building Density: 52%
Green Coverage Ratio: 35%

项目地点：中国广东省佛山市
开 发 商：信基集团
设计单位：广州市景森工程设计顾问有限公司
总（规划）用地面积：29 320 m²
总建筑面积：187 709 m²
容 积 率：4.9
建筑密度：52%
绿 化 率：35%

Features 项目亮点

The hotel facade expresses the theme by outlining an elliptical space with an arc, with the radian connects with commercial block smoothly, the natural transition endows the building much more tension. As it is the key image display surface of the project, designers injected a new element—LED lighting system under the glass curtain wall.

酒店的立面以弧线勾勒出的一个椭圆空间表达主题，随着圆角的弧度顺畅地连接商业体块，自然的过渡使建筑持有张力。作为重点表达的项目形象展示面，设计师注入了一种新的元素——玻璃幕墙下的灯光系统。

■ **Location Overview**

As a part of Guangdong urban commercial complex, the project land is located on Guangzhou-Foshan central axis and is adjacent to Nanhai High-tech Zone. Boasting a straight line distance of about 30 km both to Foshan downtown and Guangzhou downtown, it is in the heart of Guangzhou-Foshan Economoc Belt.

■ 区位概况

项目地块属于广东都市型商业综合体，坐落于广佛中轴范围上——临近南海高新技术区，到达佛山城区和广州城区的直线距离均在30 km左右，是佛经济带核心地带。

■ **Overall Planning**

The recessive square near the southern crossroad is the main entrance for high-end commercial plaza, which attracts consumers with highly flexible spatial combination and modern fashionable facade. The hotel and residences are arranged in line not only to avoid blocking view and highlight the integral facade image, but also to make more space to set the roof garden for the commercial part, which makes the commercial complex more stereoscopic and dynamic.

■ 整体规划

南部十字路口处的退让式入口广场为高端商业广场主入口，以高度灵活的平面空间组合和现代时尚的立面造型吸引商业消费人流。酒店与住宅一线展开，既避让了建筑遮挡从而强调商业立面整体化形象，又能退让更大的空间以满足屋顶花园的布置，让商业城市综合体更立体、更活跃。

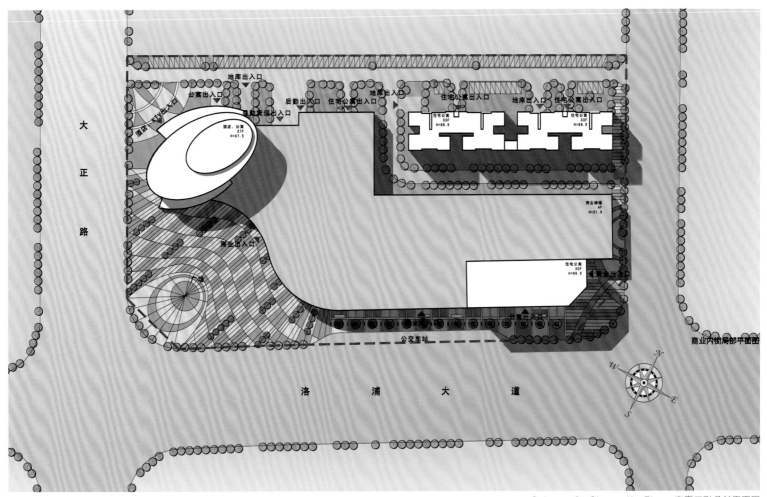

Scheme 3 Chromatic Plan 方案三彩色总平面图

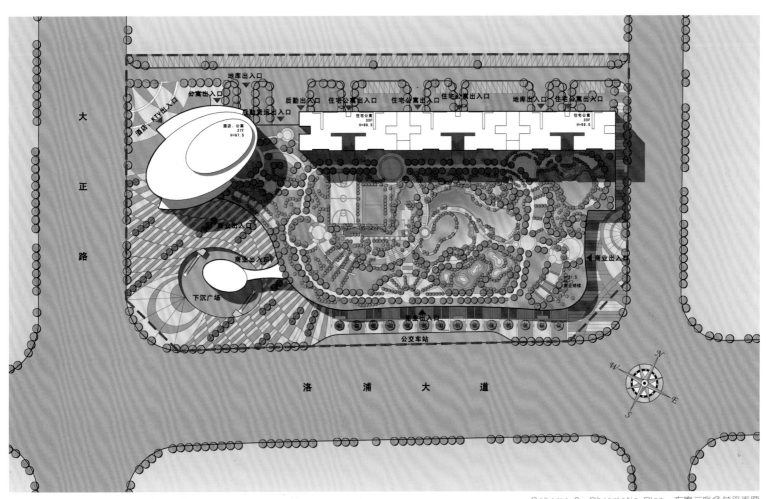

Scheme 2 Chromatic Plan 方案二彩色总平面图

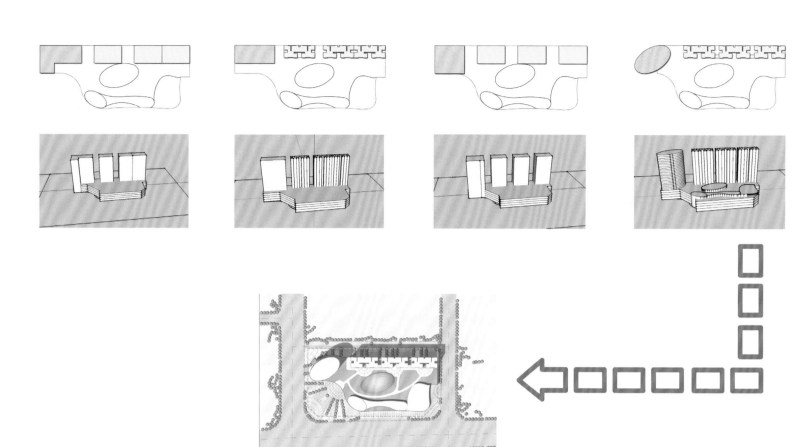

Plan Analysis 2 (Space Form Analysis)　总平面分析二（空间形态分析）

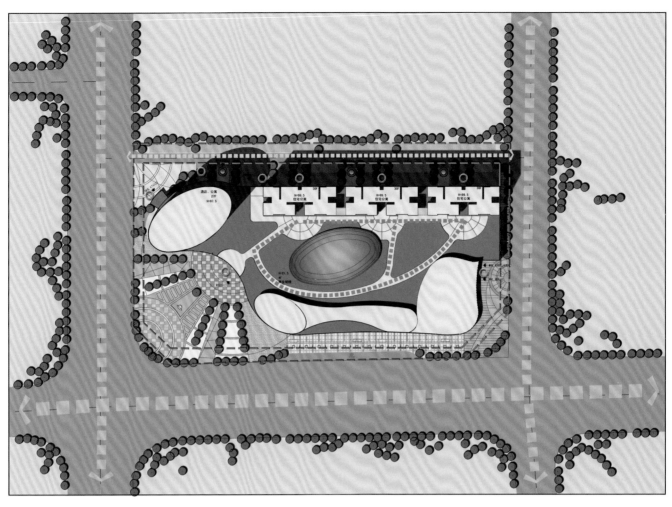

Traffic Analysis Drawing 交通分析图

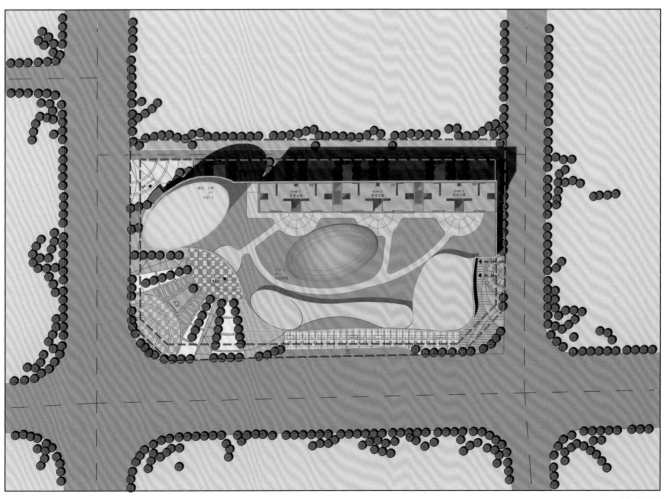

Functional Analysis Drawing 功能分析图

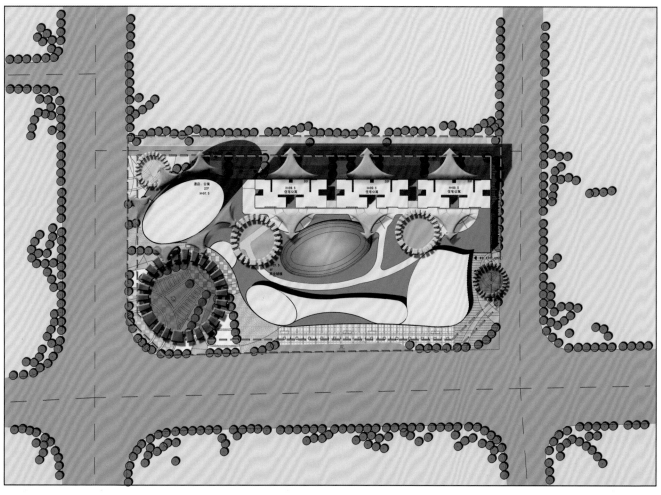

Landscape Analysis Drawing 景观分析图

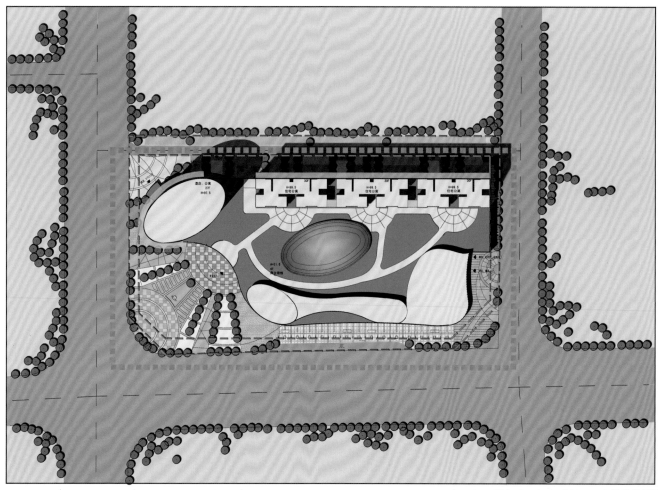

Fire Analysis Diagram 消防分析图

■ Buildings

Commercial activities are divided into two kinds: street shopfront and centralized classified business. Commercial annex has 4 floors in total. A flexible arc line runs from the main entrance through the atrium to the secondary entrance in the east. The 1st floor is dedicated to shopfront, special retail and supermarket.

Hotel & apartment share a 27-storey building. The first floor has hotel lobby and public rest area. There are two sets of three elevators. Since the management model of the hotel and apartment requires independent vertical transportation, the 5th floor is the roof floor for the annex building and the empty floor for the hotel as well. The 6th–16th floor belongs to the hotel and 17th–27th belongs to the residential part.

■ 建筑单体

商业业态分为沿街商铺与集中分类商业两类，商业裙楼共为4层。商业平面由主入口设置灵动的弧线经中庭贯穿东部次入口，首层设置沿街商铺、专业零售及超市。

酒店及公寓共享一栋27层的建筑。首层设置酒店大堂、公共休息区域。而电梯设置为两组三梯的布局，由于酒店与公寓的管理模式需独立进行竖向交通分割，第五层为裙楼屋面层，故此将该层设置为酒店架空层。6~16层为酒店部分，17~27层为住宅部分。

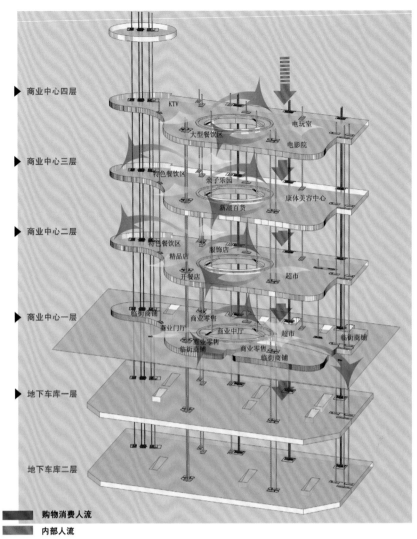

Commercial Network Analysis Drawin　商业流线分析图

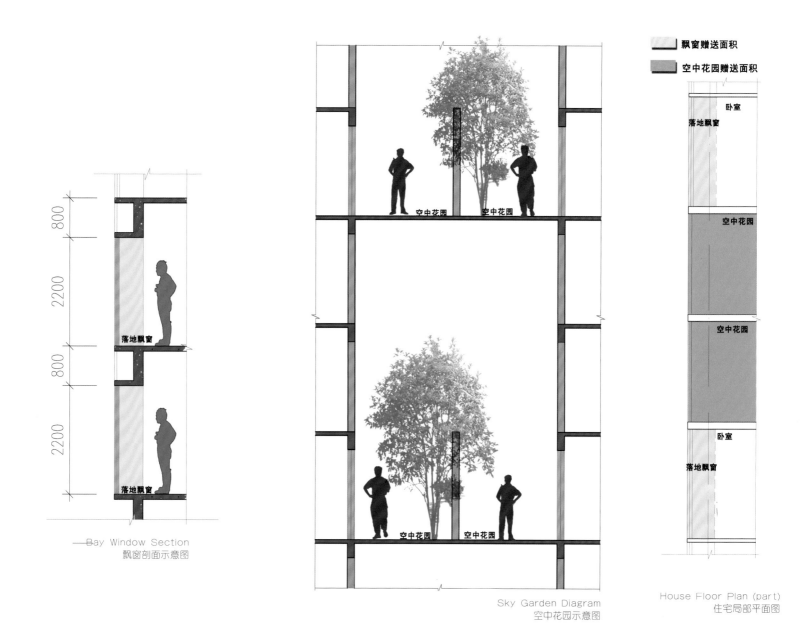

Bay Window Section
飘窗剖面示意图

Sky Garden Diagram
空中花园示意图

House Floor Plan (part)
住宅局部平面图

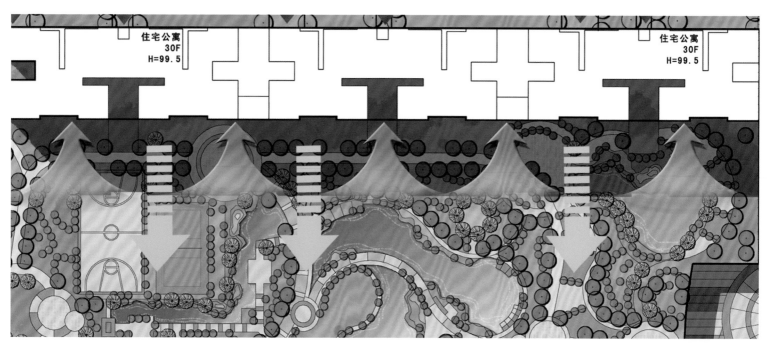

Residential Landscape Analysis Drawing 住宅景观分析图

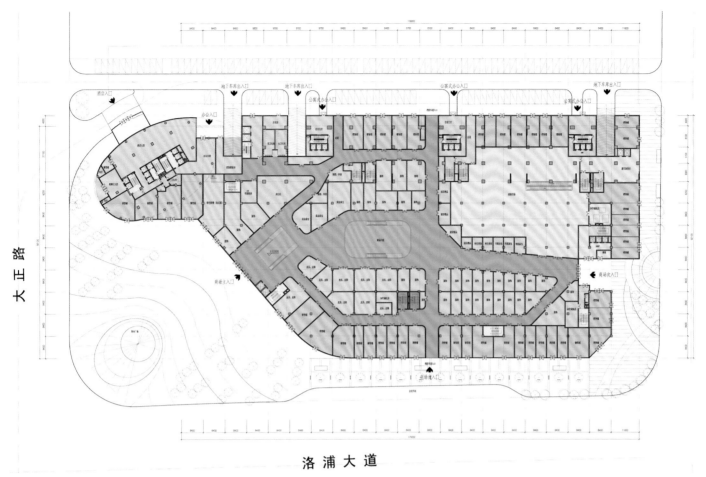

Ground Floor Plan 首层平面图

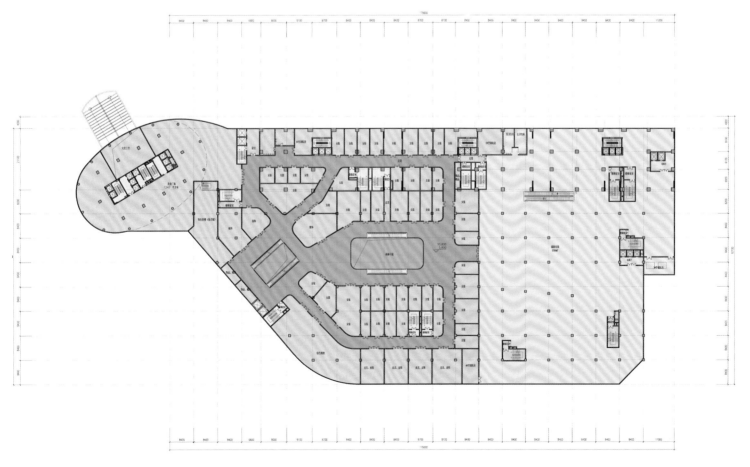

Second Floor Plan 二层平面图

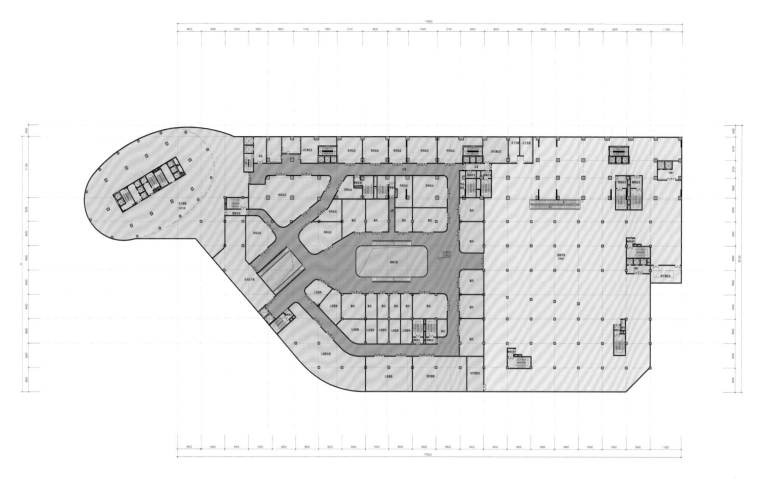

Third Floor Plan 三层平面图

Fourth Floor Plan 四层平面图

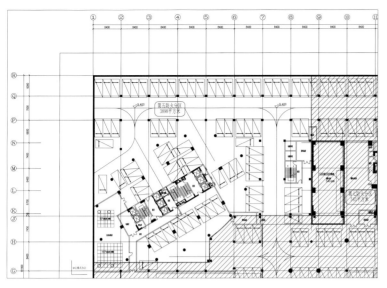
1# Hotel Office Basement 2nd Floor Plan
1# 酒店、办公地下二层平面图

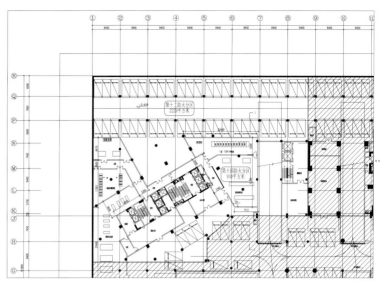
1# Hotel Office Basement 1st Floor Plan
1# 酒店、办公地下一层平面图

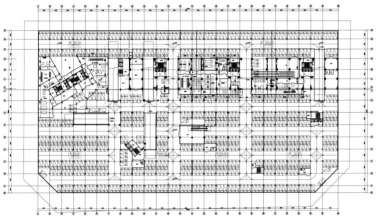
Basement Garage 1st Floor Plan
地下室车库地下一层平面图

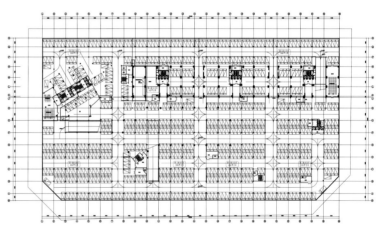
Basement Garage 2nd Floor Plan
地下室车库地下二层平面图

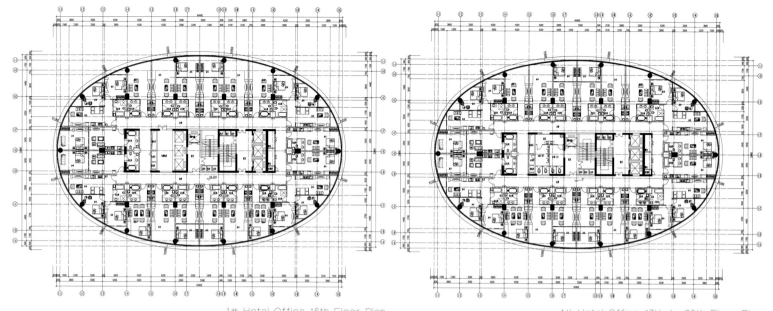

1# Hotel Office 16th Floor Plan
1# 酒店办公十六层平面图

1# Hotel Office 17th to 26th Floor Plan
1# 酒店办公十七至二十六层平面图

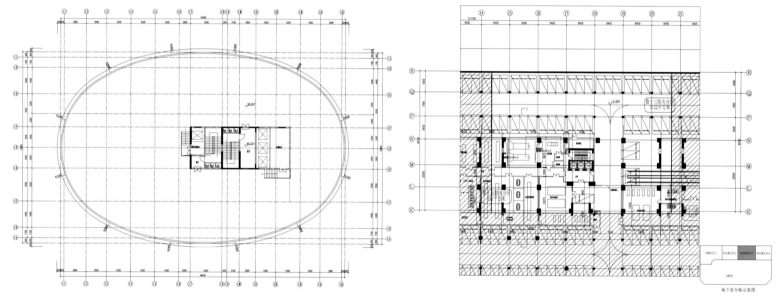

1# Hotel Office Roofing Layer Plan
1# 酒店办公屋面层平面图

3# Apartment Office Basement 1st Floor Plan
3# 公寓式办公地下一层平面图

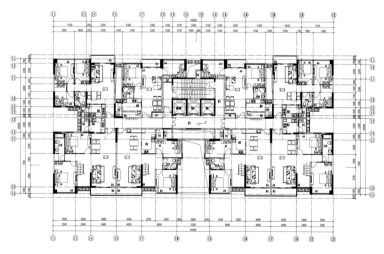

2# Apartment Office 28th Floor Plan
2# 公寓式办公二十八层平面图

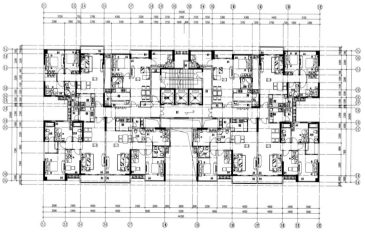

2# Apartment Office 7th to 27th Floor Plan
2# 公寓式办公七至二十七层平面图

■ Building Facade

In consideration the character of the hotel and apartment, designers pay close attention to select colors and materials. Commercial annex is characterized by modern steel form and a large curved glass wall inlaid with shutter slats, which enhance the commercial grade. The division of materials layers makes the annex more dynamic and outstanding.

The hotel facade expresses the theme by outlining an elliptical space with an arc, with the radian connects with commercial block smoothly, the natural transition endows the building much more tension. As it is the key image display surface of the project, designers injected a new element—LED lighting system under the glass curtain wall.

In residential facade design, designers use grey & white horizontal lines accordingly to echo with the buildings elements in commercial part and hotel part. In order not to make a large breadth, a sky garden is created. Glass surface interplays with the horizontal lines to express the relation of solid and void appropriately.

■ 建筑立面

结合酒店公寓及住宅的性质，色彩与材质的运用都极其讲究。商业裙楼采用现代的钢构形式和大片弧形的玻璃幕墙，其中加以百叶状装修条表达细部，提升商业档次。材质层次的划分让商业裙楼更为立体鲜明。

酒店的立面表达主题以弧线勾勒一个椭圆的空间，随着圆角的弧度顺畅地连接商业体块，自然的过渡使建筑持有张力，作为重点表达的项目形象展示面，设计师注入了一种新的元素—玻璃幕墙下的LED灯光系统。

在住宅立面上，设计师应运而生依据商业与酒店的整体性以灰白粗壮的横向线条构成其相互呼应的建筑元素。为了不使住宅的面宽界面过长，在局部做了空中花园的处理，让其突显灵动。玻璃介面在实体线条内让虚实关系表达非常到位。

Changsha Shimao Plaza
长沙世茂广场

Keywords 关键词

Curved Streamline
曲线弧形

Natural Elements
自然符号

Glass Curtain Wall
玻璃幕墙

Features 项目亮点

The commercial podium conveys strong and intense commercial atmosphere and the bending arc adds stronger sense of levels, which is reminiscent of the mountains upon mountains, grand and momentous.

商业裙楼立面富有变化，商业气息浓厚，弯曲的弧形流线使建筑更具层次感，犹如层层叠叠的山峰，给人以磅礴之势。

Location: Changsha, Hunan, China
Developer: Changsha Shimao Investment Co.,Ltd.
Architectural Design: Shanghai Zhongjian Architectural Design Institute Co., Ltd.
Collaborated With: The Jerde Partnership
Total Site Area: 16,073.14 m^2
Total Floor Area: 229,156.68 m^2
Plot Ratio: 12.3%
Green Coverage Ratio: 25%

项目地点：中国湖南省长沙市
开 发 商：长沙世茂投资有限公司
建筑设计：上海中建建筑设计院有限公司
合作建筑设计：美国捷得国际建筑师事务所
总用地面积：16 073.14 m^2
总建筑面积：229 156.68 m^2
容 积 率：12.30%
绿 地 率：25%

■ Overview

The project is located in Zhaojiaping of Lotus District in Changsha, facing Jianxiang South Road and intersecting with Wuyi Road. Wuyi Road leads to Changsha Railway Station in the east and Xiangjiang River First Bridge in the west while Jianxiang South Road leads all the way to Lotus Middle Street in the south. The project neighbors Phoenix Street on the east, Panhou Street on the south and old downtown buildings on southeast. The site is also near to Lotus Square, Martyr Memorial Park, Hunan Provincial Museum and Helong Sports and Leisure Plaza, etc. Covering a total land area of 16,073.14 m^2, of which 13,827.34 m^2 are occupied, the plaza stands 75 floors high and has 4 underground floors. The site enjoys superior location and development potential with complete living facilities and convenient transportation in the surroundings. The project will be forged into a dynamic, international, modern and fashionable landmark building.

■ 项目概况

项目位于长沙芙蓉区肇家坪，面临建湘南路，与五一大道相交，沿五一大道向东直通长沙火车站、向西通向湘江一桥，沿建湘南路向南直达芙蓉中路。东侧有凤凰街，南侧有潘后街，东南侧为老城区建筑。基地临近芙蓉广场，且烈士公园、湖南省博物馆、贺龙体育休闲娱乐广场等遍布办公区周围，地势较平坦。基地总用地面积为16 073.14 m^2，净用地为13 827.34 m^2，地上为75层，地下为4层。周边生活配套设施完善，交通便利，地理位置优越，极具发展潜力，力图创造一个富有活力、具有国际化理念、现代时尚的标志性建筑。

■ Planning

The project consists of a commercial podium and an office tower, forming a graceful curve connecting rightly with the podium of the neighboring building. The main entrance is located in northeast direction, which is planned to be a spacious square to accommodate more and provide larger vision and views. The main entrance of office building is located in the west and northeast direction, which enjoys large green space and square space to effectively broaden the entrance area and increase ecological part. In the east direction, there sets up an independent entrance for catering to fulfill the function of the project. The circulation of the project is clear and well organized.

Site Plan 总平面图

■ 规划布局

建筑由商业裙楼和办公塔楼两部分组成。整个建筑形体成一柔美的曲线弧形，与旁边地块的裙楼完美对接。商业主入口位于东北侧，入口位置规划为广阔的入口广场，使其容纳性更大、视野更广阔，并形成了很好的景观视野。办公的主入口位于西侧及东北侧，有较大的绿化空间及广场空间，使其入口更加开阔且增加了生态区。在东侧还设有餐饮的单独出入口，使其功能更加合理。交通流线组织亦清晰明了，布局合理。

Architectural Design

The entire space is simple, grand, modern and characteristic. In design process, this project gives much attention to the outside environment of the city and contextual design. The entire building possesses abundant architectural contour lines, coordinated architectural syntagmatic relation, organic mass combination and strong rhythm and tempo in light and shadow changing. The architectural style is elegant, dignified, clear, bright and attractive, creating a complete body and atmosphere.

The project is enlightened by Chinese traditional paintings. With valley and high mountain waterfall as the main concepts, the elevation applies natural elements on the projects. On the side along Jianxiang Road, a five-floor high commercial podium building with expansionary force is fixed on Site A, which forms a spacious circular arc with wide horizon.

The commercial podium building conveys strong and intense commercial atmosphere and the bending arc adds stronger sense of levels, which is reminiscent of the mountains upon mountains, grand and momentous. The elevation design for tower building adopts "high waterfalls" of glass curtain to form innervation of water running down with the reflection of glass and LED lights and gather in the middle of the square. Following the water down, the waterfalls pour on each level to give birth to gardens in the air on different levels.

■ 建筑设计

整个建筑简洁、大气、现代、个性鲜明。本方案在创作时强调了城市的外部空间环境和脉络设计，整个建筑轮廓线丰富，建筑空间形态组合关系有机协调，建筑的有机体量组合及其光影变化具有强烈的节奏感和韵律感，建筑风格端庄大方、高雅脱俗、清朗简洁且新颖夺目，创造了完整的形体和气势。

项目受到传统中国画特性启发，以"峡谷"和"高山瀑布"为主要设计理念，立面设计多采用自然符号。基地沿建湘路的主要沿街面，一座5层高充满扩张力的弧形商业裙楼有力地稳固在地块A，使空间形成一个视野开阔的圆弧形广场。

商业裙楼立面富有变化，商业气息浓厚，弯曲的弧形流线使建筑更具层次感，犹如层层叠叠的山峰，给人以磅礴之势。塔楼的立面设计采用玻璃幕的"高山瀑布"，在玻璃和LED照明反射下融合出一种水往下流淌的动感，汇聚在中央广场中。顺着水流而下，瀑布浇注在每一层楼中，形成了层层的空中花园。

■ Landscape Design

Landscape design covers road greening, commercial plaza design and landscape architectures. Road greening is carried out with evergreen trees, sidewalk and plaza stones. The commercial plaza applies hard pavement in combination with framed flower bed to increase the dynamism of space. Meanwhile, activity ground, exercising ground and art gallery are also equipped in the place.

■ 景观设计

景观设计包括：道路绿化、商业广场和景观小品。道路绿化采用常绿树装点，人行道以及广场铺贴地砖等方式。商业广场采用硬质铺装与景观花池绿化相结合的手法，增强空间的活跃感。同时还布置了活动场地、健身场地和艺术廊等。

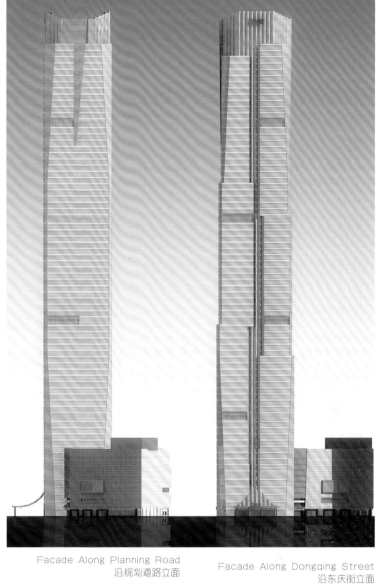

Facade Along Yunjia Lane
沿允嘉巷立面

Facade Along Jianxiang South Road
沿建湘南路立面

Facade Along Planning Road
沿规划道路立面

Facade Along Dongqing Street
沿东庆街立面

Jinan Dinghao Plaza
济南丁豪广场

Keywords 关键词

L Shape Layout
L 型布局

Peeling and Cutting Disposition
削切处理

Free Combination
自由组合

Location: Jinan, Shandong, China
Developer: Dinghao Group
Architectural Design: ACBI

项目地点：中国山东省济南市
开 发 商：丁豪集团
设计单位：加拿大宝佳国际建筑师有限公司

Features 项目亮点

Commercial street facade is a combination of wood louvers, perforated metal panels and glass. A contrast is set between the interior facade and the exterior facade, which increase its activeness. Besides, the space is also blurred to make more fun.

商业街建筑外立面材料采用木质百叶、金属穿孔板和玻璃相结合。内外立面进行了反差设计，既增加了立面的活跃性，同时将空间进行了模糊化处理，增加了空间趣味性。

■ Overview

Located in the core area of High-tech Zone, Dinghao Plaza is adjacent to Jinan International Convention and Exhibition Center and Jinan High-tech Zone Administration Center. It is positioned as a large-scale city complex which consists of commerce, conference, entertainment and shopping, coinciding with the regional planning public center in the development zone. It accommodates five-star hotel, comprehensive commercial pedestrian street, apartment and commercial office building that provide a relatively complete industrial chain.

■ 项目概况

项目位于高新区中心区核心位置，紧邻济南国际会展中心及济南高新区行政审批服务中心。项目定位为集商务、会议、休闲和购物于一身的大型城市综合体，与开发区规划中的地区级公共中心相吻合。项目内容涵盖五星级酒店、综合性步行商业街、公寓及商务办公楼，为地区及公共中心的打造提供了相对完整的产业链模式。

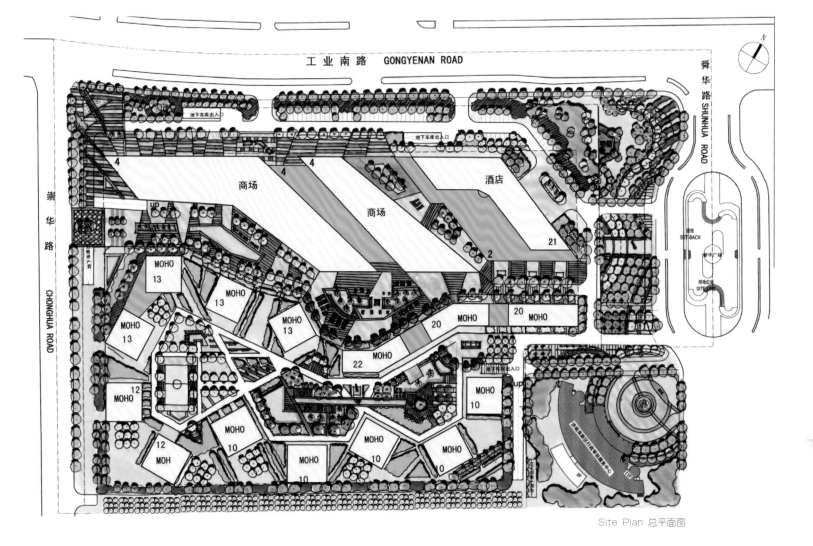

Site Plan 总平面图

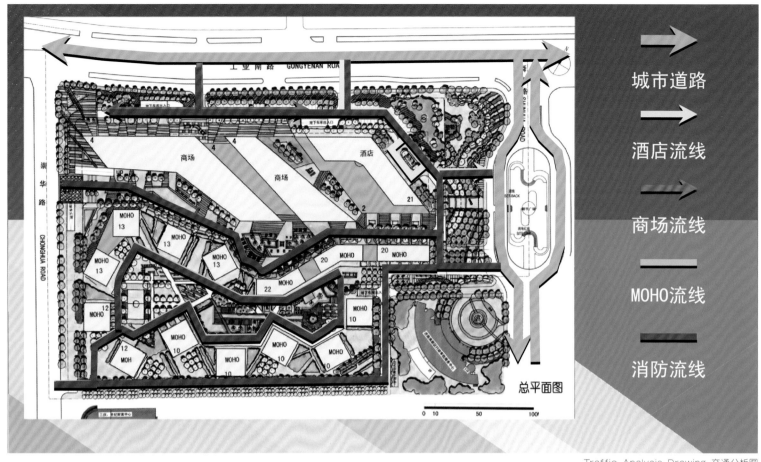

Traffic Analysis Drawing 交通分析图

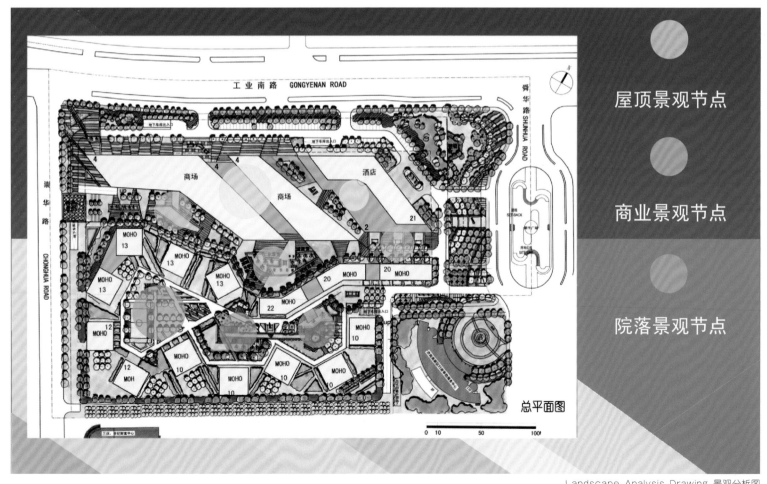

Landscape Analysis Drawing 景观分析图

East Elevation 东立面图

North Elevation 北立面图

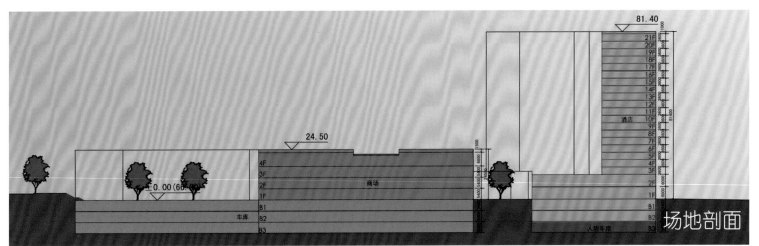

Section 1-1 剖面图 1-1

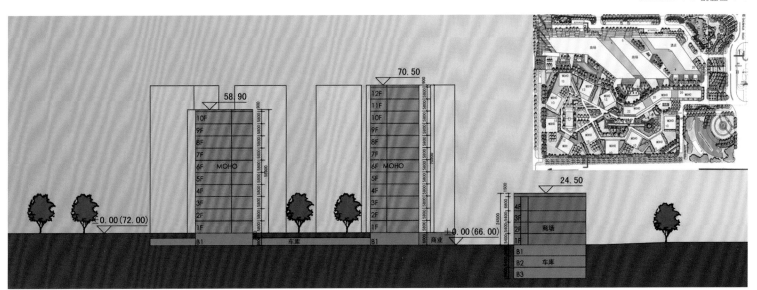

Section 2-2 剖面图 2-2

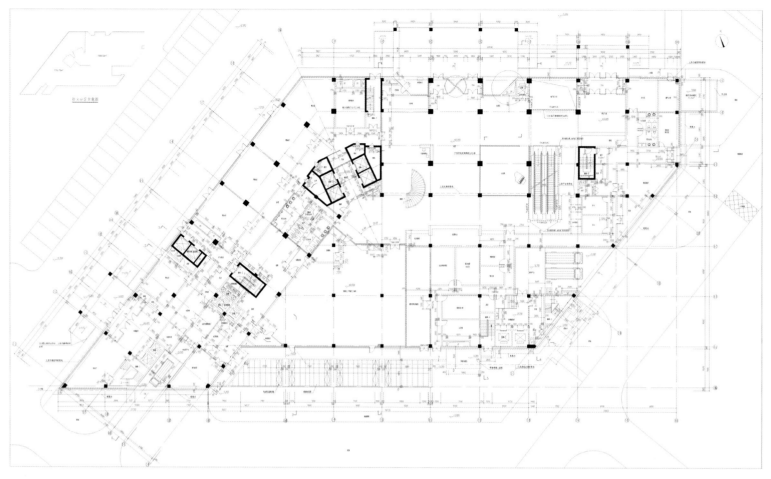

Ground Floor Plan of Hotel 酒店首层平面图

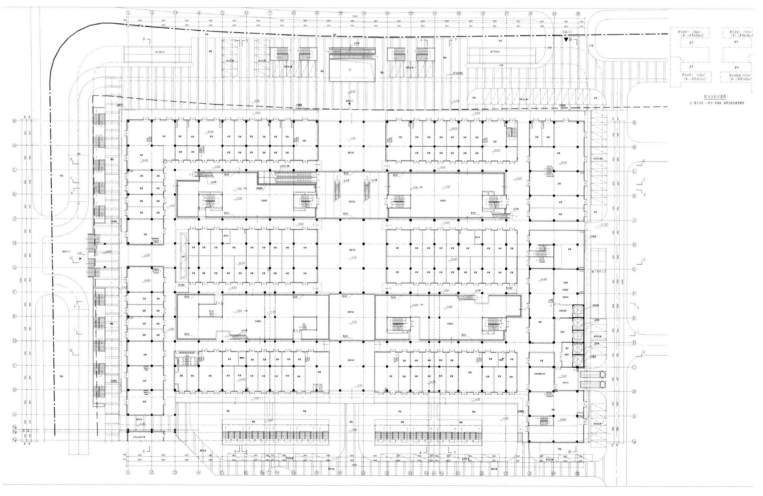

Ground Floor Plan of Market 商场首层平面图

■ Architectural Design

The 80m high five-star hotel occupies the northeast corner of the land, echoes with Jinan International Convention and Exhibition Center and Jinan High-tech Zone Administration Center to form a landmark.

Comprehensive commercial pedestrian street extends along Gongye South Road and intersects with Chonghua Road. Its image is maximized. The front square can provide sufficient space to distribute shopping crowd and vehicles, and the internal road reduces the usual traffic pressure in commercial building efficiently by linking two main city roads.

Commercial office building is on the south side of the commercial street, and a commercial courtyard and an organic whole is formed by combining the shopfront on the ground floor and the commercial street. The office people enter the building through a different entrance which separates them with commercial crowd effectively, thus to increase the independence of commercial activities and reduce interference.

Apartment is located in the southernmost of the land. Through vertically processing the height difference, three-dimensional transportation is used to separate people flow and vehicle flow. And an independent courtyard is provided for the residents to have leisure activities.

■ 建筑设计

　　五星级酒店占据了东北角，建筑高度为80 m，与济南国际会展中心和济南高新区行政审批服务中心相呼应形成区域标志性建筑。

　　综合性步行商业街沿工业南路和崇华路交口布置，形象得到最大化展示。前广场可以为购物人群及车辆交通提供充足的集散空间，并通过与两条城市主路相连的内部道路有力地解决商业建筑常有的交通压力过大问题。

　　商务办公楼布置在商业街南侧，底层商业与商业街组成商业院落，形成有机整体。办公人群通过不同入口进入，与商业人群有效分离，增加业态的独立性，减少干扰。

　　公寓位于整个用地最南部，通过对用地高差的竖向处理，采用立体交通的模式，对人车进行分流，并提供独立庭院供人们进行休闲活动。

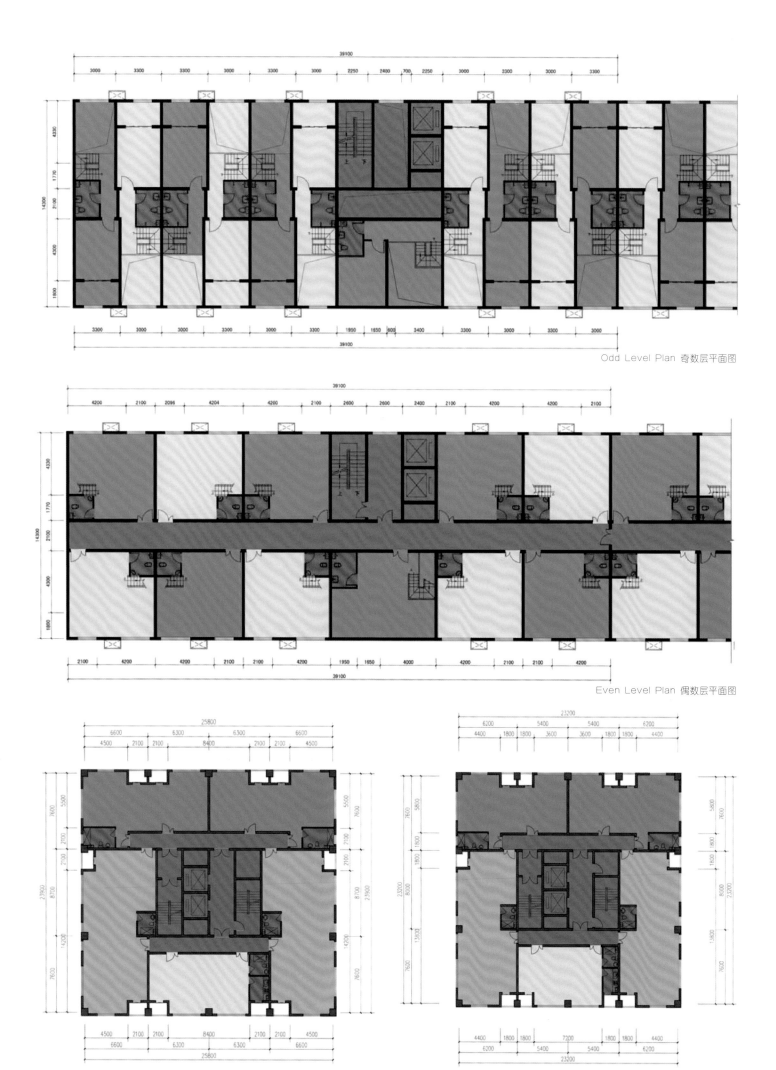

Odd Level Plan 奇数层平面图

Even Level Plan 偶数层平面图

Tower Standard Layer Plan 1 塔式标准层平面图 1

Tower Standard Layer Plan 2 塔式标准层平面图 2

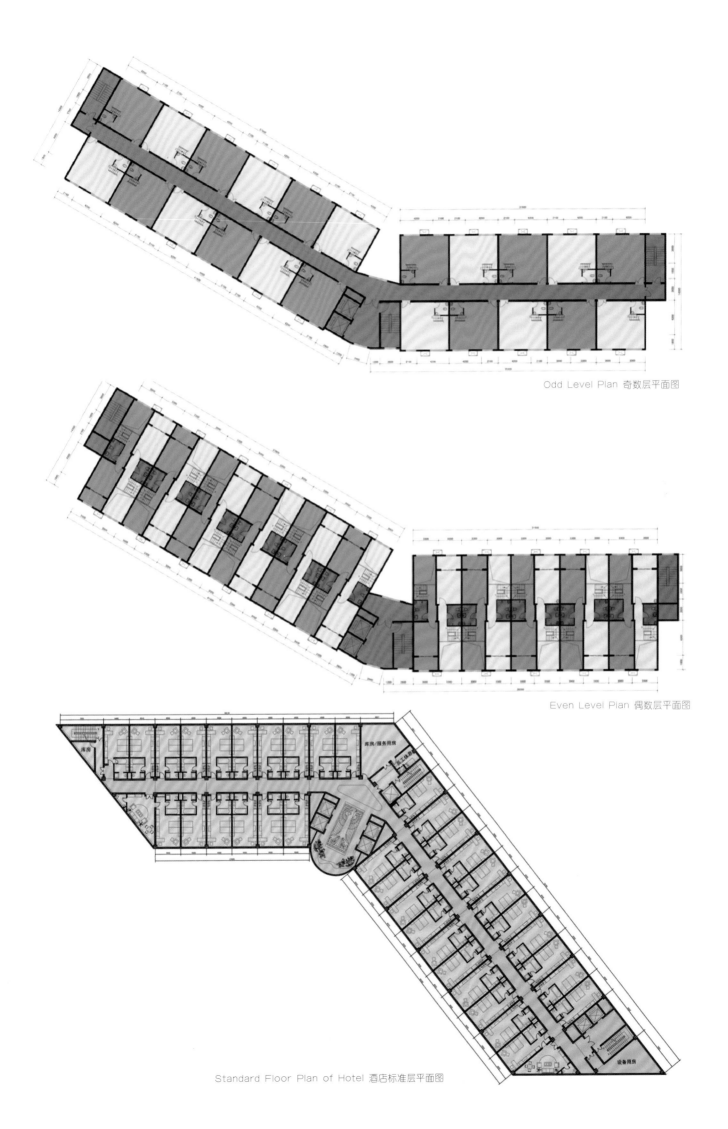

Odd Level Plan 奇数层平面图

Even Level Plan 偶数层平面图

Standard Floor Plan of Hotel 酒店标准层平面图

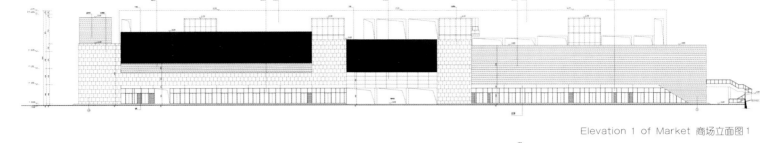

Elevation 1 of Market 商场立面图1

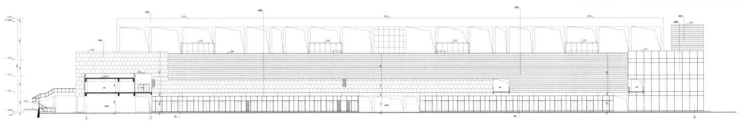

Elevation 2 of Market 商场立面图2

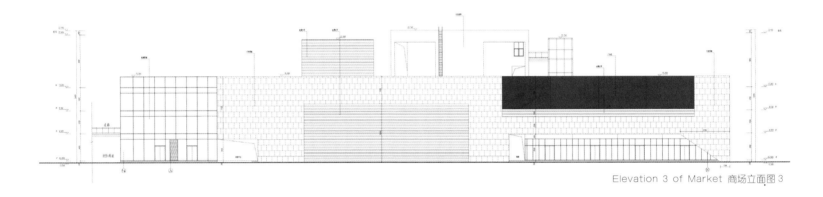

Elevation 3 of Market 商场立面图3

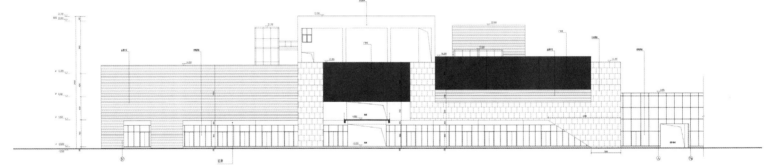

Elevation 4 of Market 商场立面图4

Elevation 1 of Hotel 酒店立面图1

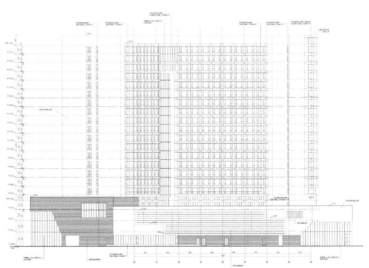

Elevation 2 of Hotel 酒店立面图2

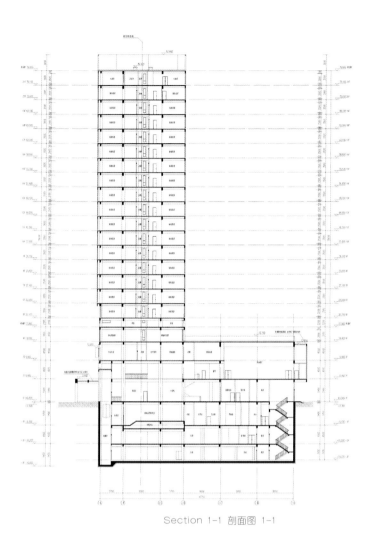

Section 1-1 剖面图 1-1

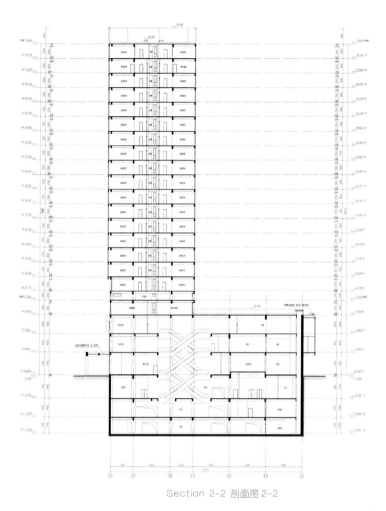

Section 2-2 剖面图 2-2

■ Fcade Design

Hotel facade is mainly composed of stone, metal components, glass, wood and concrete structures. The simple and generous style highlights the elegance of the hotel.

Commercial street facade is a combination of wood louvers, perforated metal panels and glass. A contrast is set between the interior facade and the exterior facade, which increase its activeness. Besides, the space is also blurred to make more fun.

Commercial office building shapes a sharp contrast with apartment in color, emphasizing vertical sense to enhance visual impact.

■ 立面设计

酒店立面材料主要采用石材、金属构件、玻璃及部分木质和混凝土构件。立面风格简洁大气，以突出酒店高贵气质。

商业街建筑外立面材料采用木质百叶、金属穿孔板和玻璃相结合。内外立面进行了反差设计，既增加了立面的活跃性，同时将空间进行了模糊化处理，增加了空间趣味性。

商务办公楼与公寓采用色彩强烈对比的手法，强调建筑的竖向线条感，以增强建筑的视觉冲击力。

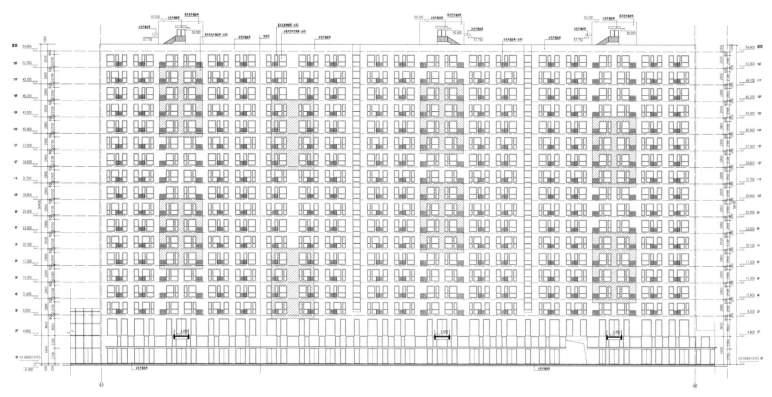

Elevation 1 of Apartment and Office Building 公寓办公立面图1

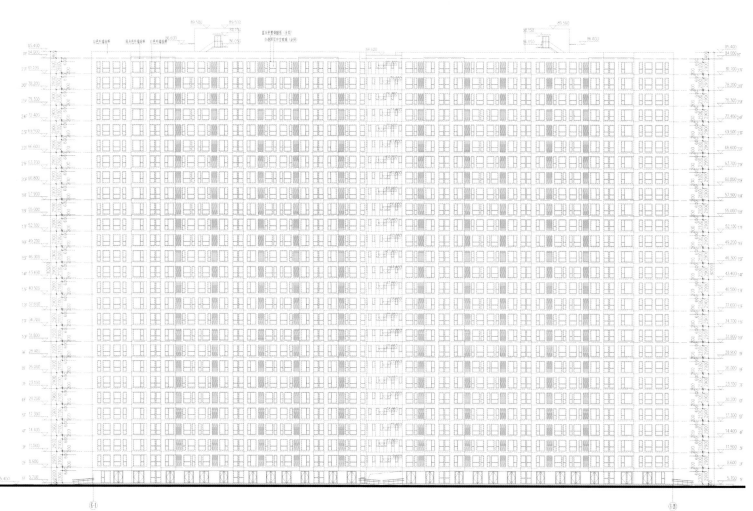

Elevation 2 of Apartment and Office Building 公寓办公立面图2

Elevation 3 of Apartment and Office Building 公寓办公立面图 3

Business Street in Huishan District, Wuxi
无锡惠山区商业街

Keywords 关键词

Classic Small Town
古典小镇

Organic Integration
有机融合

Dumbbell-shaped Layout
"哑铃型"布局

Features 项目亮点

The concept for this long development is one of the organic penetration theorys, in which each plot links together, organically, like cells, punctuated at each end with corner-capturing architectural features, distributing the stream of people well to maximize commercial value.

街区整个规划通过有机渗透的长效发展理论，将每个地块有机地联系到一起，通过"哑铃型"的商业布置和街角引入，将商业街人流均匀布置，使得商业价值最大化。

Location: Wuxi, Jiangsu, China
Architectural Design: KAZIA.LI Design Collaborative
Architects: Ryan Moss, Jason Gao
Developer: Wuxi City Imperial Real Estate Development Co., Ltd.
Size: 63,070 m²
Total Site Area: 42,071 m²
Building Density: 44.4%
Plot Ratio: 1.5
Green Coverage Ratio: 19%

项目地点：中国江苏省无锡市
设计单位：天津凯佳李建筑设计事务所
设 计 师：瑞恩·莫斯、高昇
开 发 商：无锡市御房地产开发有限公司
建设规模：63 070 m²
总用地面积：42 071 m²
建筑密度：44.4%
容 积 率：1.5
绿 化 率：19%

■ **Overview**

Situated in the educational park in Huishan District, Wuxi, this project for commercial use is located to the east of Xin'ou Road and by the two sides of Weishan Road. The overall design of this region fits the market development trend and highlights the sharing of resources and optimization of ecological environment. It intended to provide leisure, entertainment, and enrichment opportunities for the students of the area.

■ 街区概况

该项目位于无锡惠山区藕塘职教园区内，新藕路以东、纬三路两侧，地块用地性质属商业配套用地。根据项目所处区域的总体发展方向，适应市场发展潮流，强调环境资源共享和生态优化的设计前提下，本方案拟在该基地内建造以学生街为主题的商业环境，为周边大学生提供娱乐休闲活动，丰富日常业余生活。

■ **Block Plan**

This project is a combination of five plots. According to the site's location and the surrounding resources, it was designed to be a middle and high-end commercial block to service the students and community residents around. The concept for this long development is one of the organic penetration theorys, in which each plot links together, organically, like cells, punctuated at each end with corner-capturing architectural features, distributing the stream of people well to maximize commercial value.

■ 街区规划

本项目共有五个地块。根据地块位置特殊性及周边资源，努力将项目打造成以服务周边大中专院校的学生日常生活消费的中高端商业为主，吸引项目周边部分社区居民消费的商业街区。整个规划通过有机渗透的长效发展理论，将每个地块有机地联系到一起。通过"哑铃型"的商业布置和街角引入，将商业街人流均匀布置，使得商业价值最大化。

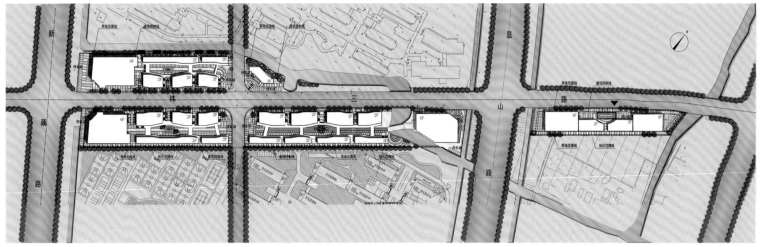

Site Plan 总平面图

■ Block Design

In terms of facade treatment, designers knew well about classical building and simplified it to present a classical small town style while retaining the essence of modern building. Amiable sense of dimension has increased the commercial quality dramatically.

■ 街区设计

在立面处理上通过对古典建筑的理解和简化，既呈现了古典小镇的风情，又不失现代建筑的精髓。尺度的亲切感使得商业品质也大幅提升。

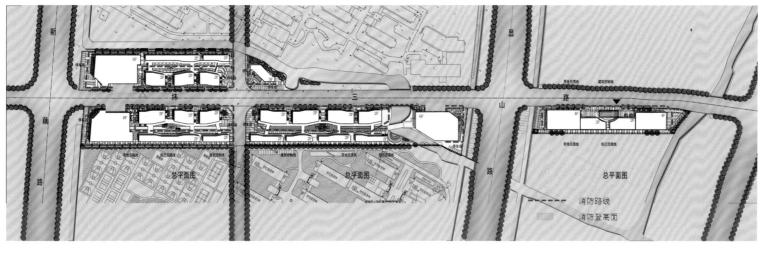
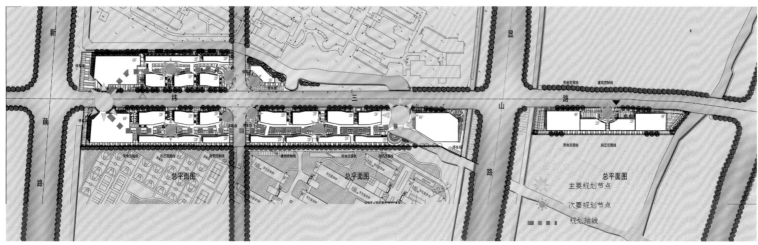

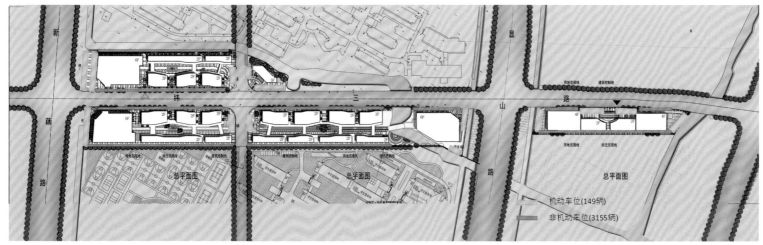

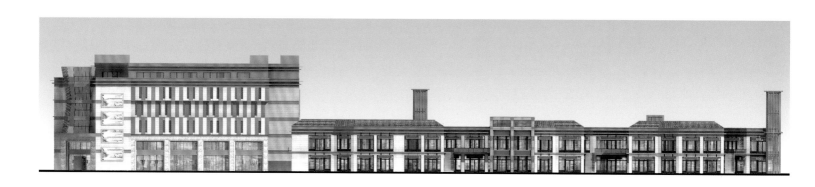

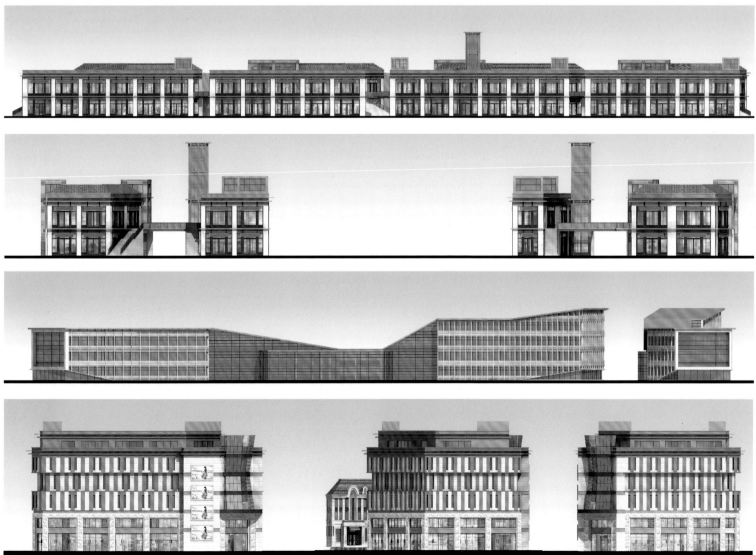

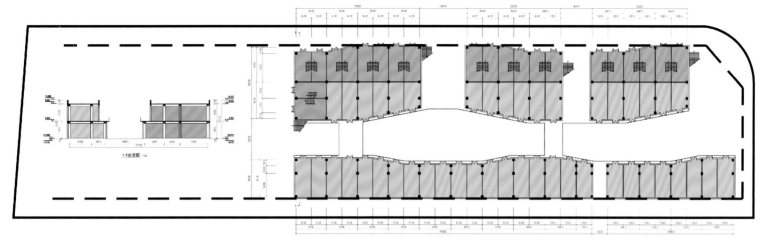

First Floor Plan 一层平面图

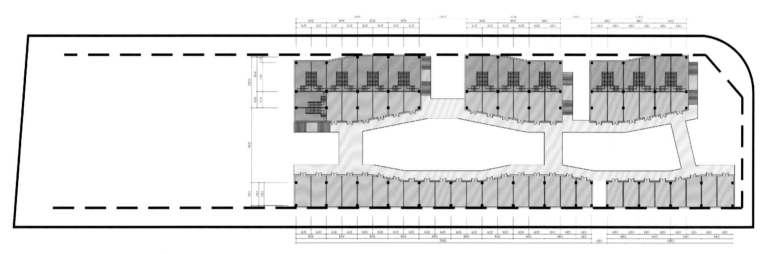

Second Floor Plan 二层平面图

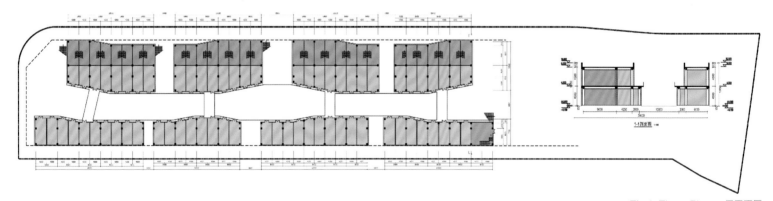

First Floor Plan 一层平面图

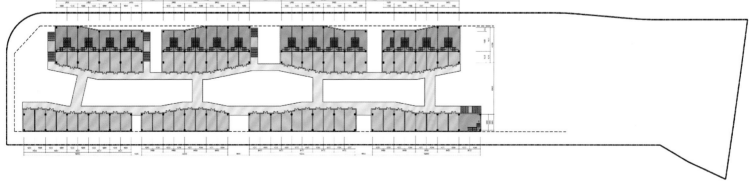

Second Floor Plan 二层平面图

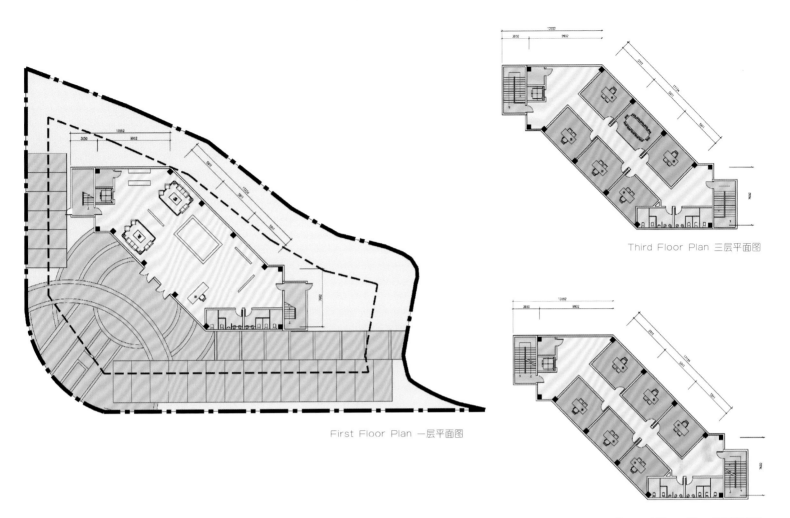

First Floor Plan 一层平面图

Third Floor Plan 三层平面图

Second Floor Plan 二层平面图

Yingtian Road Market Center in Nanjing
南京应天大街营房仓库商业开发

Keywords 关键词

Orderly Space
空间错落

Chinese Republic Architectural Style
民国风貌

Fashionable Street
时尚街区

Features 项目亮点

Focusing on the favorable orderly space, designers highly integrated the city road and surrounding buildings in terms of overall layout to express the design philosophy and design orientation. The new business and entertainment district expresses a modern vision of Chinese Republic architecture style, with garden houses grouped around a public plaza, and supporting high-end restaurants and cultural entertainment clubs.

本规划总体布局着眼于良好的空间错落关系，通过与城市道路和周边建筑类型的高度结合，表达出设计的理念和方向。项目定位在民国风格别墅时尚街区，并辅以高档品牌餐饮、文化娱乐会所作为配套。

Location: Nanjing, Jiang Su, China
Architectural Design: KAZIA.LI Design Collaborative
Architect: Han Xu
Developer: Jiangsu Heshanxin Investment Company
Total Site Area: 17,219 m²
Total Building Area: 43,784 m²
Plot Ratio: 1.69
Building Density: 45%
Green Coverage Ratio: 30%

项目地点：中国江苏省南京市
设 计 公 司：天津凯佳李建筑设计事务所
设 计 师：韩旭
开 发 商：江苏赫山鑫投资有限公司
总用地面积：17 219 m²
总建筑面积：43 784 m²
容 积 率：1.69
建筑密度：45%
绿 化 率：30%

■ **Overview**

Nestled on 17,000 m² of land in a prime location near the Hexi CBD Center, New Town Science and Technology Park, and an upscale residential district, this project is poised in an ideal location to serve its neighbors with high-end dining, culture & entertainment and club, Nanjing is a historical city with profound culture. In correspond to city culture, the new business and entertainment district expresses a modern vision of Chinese Republic architecture style, with garden houses grouped around a public plaza, and supporting high-end restaurants and cultural entertainment clubs.

■ **街区概况**

江苏省军区后勤部应天大街营房仓库地块区占地面积约17 000 m²，坐拥河西CBD中心、新城科技园及高档住宅小区，是开发高档餐饮、文化娱乐、企业会所的理想之处。南京历史悠久，文化底蕴厚重。为了契合城市文化，项目定位在民国风格别墅时尚街区，并辅以高档品牌餐饮、文化娱乐会所作为配套。

■ **Street Design**

Focusing on the favorable orderly space, designers highly integrated the city road and surrounding buildings in terms of overall layout to express the design philosophy and design orientation. The over ground part in this commercial area is for 12 buildings and the underground is for basement, parking space, property management room, auxiliary equipment room, etc. And the street in Chinese Republic architecture style was equipped with complete grid road system, neighborhood and green space system. Ingenious garden houses scattered in the commercial plaza, unified but unique building groups boast an elegant place among the tall buildings, all of which form a characteristic architectural style.

■ **街区规划**

本规划总体布局着眼于良好的空间错落关系，通过与城市道路和周边建筑类型的高度结合，表达出设计的理念和方向。商业区内地上规划建设12栋建筑，地下用于建筑地下室、停车场、物业管理用房以及设备辅助用房。商业区着力打造民国建筑风格，有着完整的道路、街坊和绿地系统。道路呈格网布置，还原民国风貌。构思巧妙的花园洋房散落在商业广场，建筑群体既统一又独具特色，在河西高楼林立中获取一片优雅之地，形成独特的建筑风格。

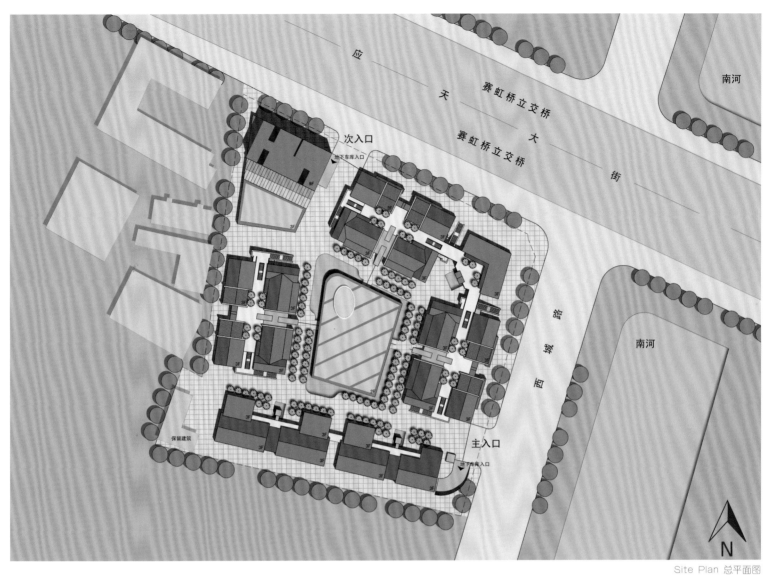
Site Plan 总平面图

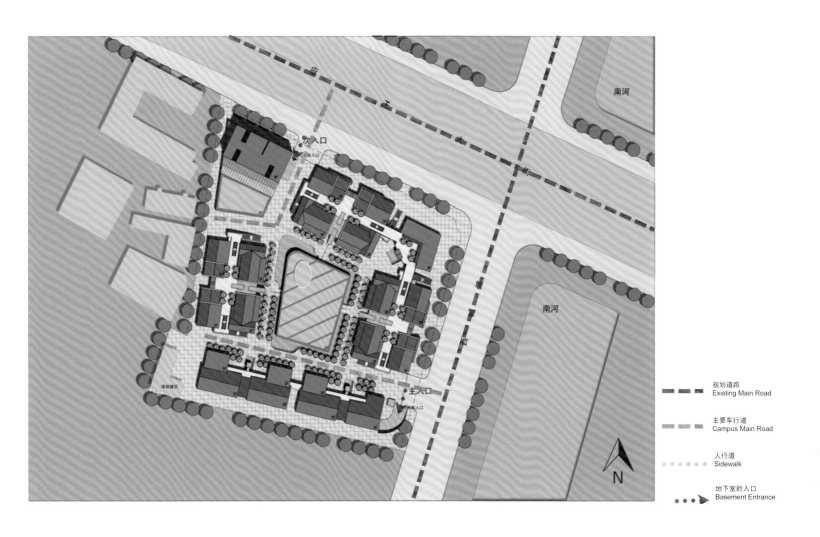
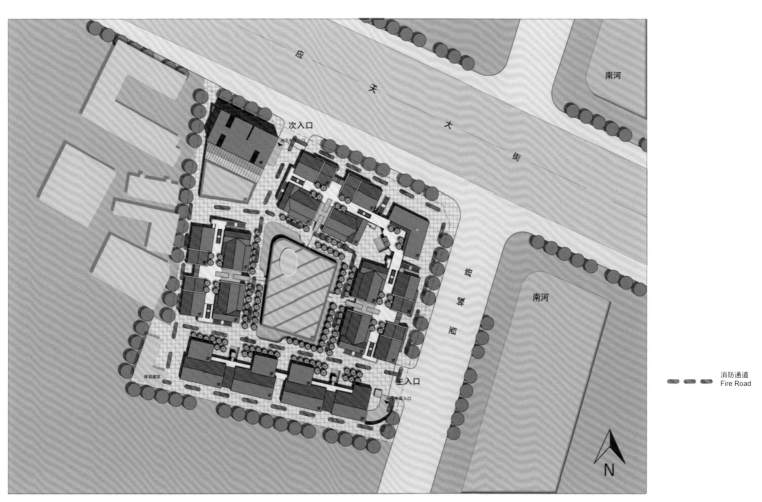

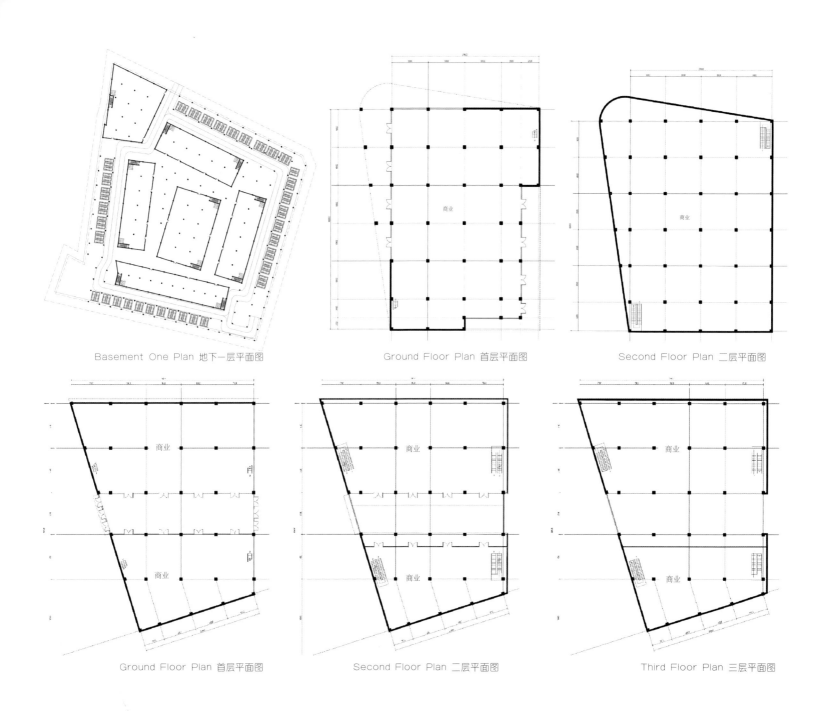

Basement One Plan 地下一层平面图　　Ground Floor Plan 首层平面图　　Second Floor Plan 二层平面图

Ground Floor Plan 首层平面图　　Second Floor Plan 二层平面图　　Third Floor Plan 三层平面图

Urban Complex
城市综合体

- Brand Strategy
  品牌策略

- Superior Location
  区位优势

- Organic Integration
  有机整合

- Unique Style
  独特风格

Core Casa, Tianjin
天津滨海新城

Keywords 关键词

- Central Courtyard 中空庭园
- Space Shaping 场所塑造
- Facade Material 立面材质

Features 项目亮点

To comply with the development of modern economy and technology, the facades of the buildings are designed with a variety of materials, presenting varied textures as well as beautiful and tall buildings.

为印证当代经济技术发展特征，建筑外立面采用多种材质，呈现出多种肌理，形成亮丽而又挺拔的建筑形态。

Location: Tanggu District, Tianjin, China
Developer: Tianjin Core Casa's Construction & Development Co., Ltd.
Architectural Design: ZPLUS
Architect: Liu Shunxiao
Total Floor Area: 800,000 m²

项目地点：中国天津市塘沽区
开 发 商：滨海新城建设发展公司
建筑设计：ZPLUS普瑞思建筑规划设计公司
总设计师：刘顺校
总建筑面积：800 000 m²

■ **Overview**

Located to the south of the toll station of Beijing-Tianjin-Tanggu Highway, and to the west of the central ring road, Core Casa Tianjin boasts a total floor area of 800,000 m², of which 580,000 m² is over ground, and 220,000 m² is underground. The main entrances are set on the east and north side.

■ **项目概况**

滨海新城位于天津京津塘高速公路收费站以南、中环线以西，地上建筑面积约为58万m²，地下建筑面积约为22万m²，共80万m²。该项目交通主要出入口方向位于东侧和北侧。

■ **Planning Concept**

The planning pays attention to creating sequential spaces, cares for people's behaviors, and focuses on the convenience of functions. It aims to provide people with a harmonious, multi-functional, safe and comfortable environment which is full of art atmosphere.

■ **规划理念**

规划中注重连续空间的塑造，关照人的行为及空间使用的便利，以丰富和谐的环境为人们提供和谐融恰、功能齐全、安全舒适且充满艺术气息的场所。

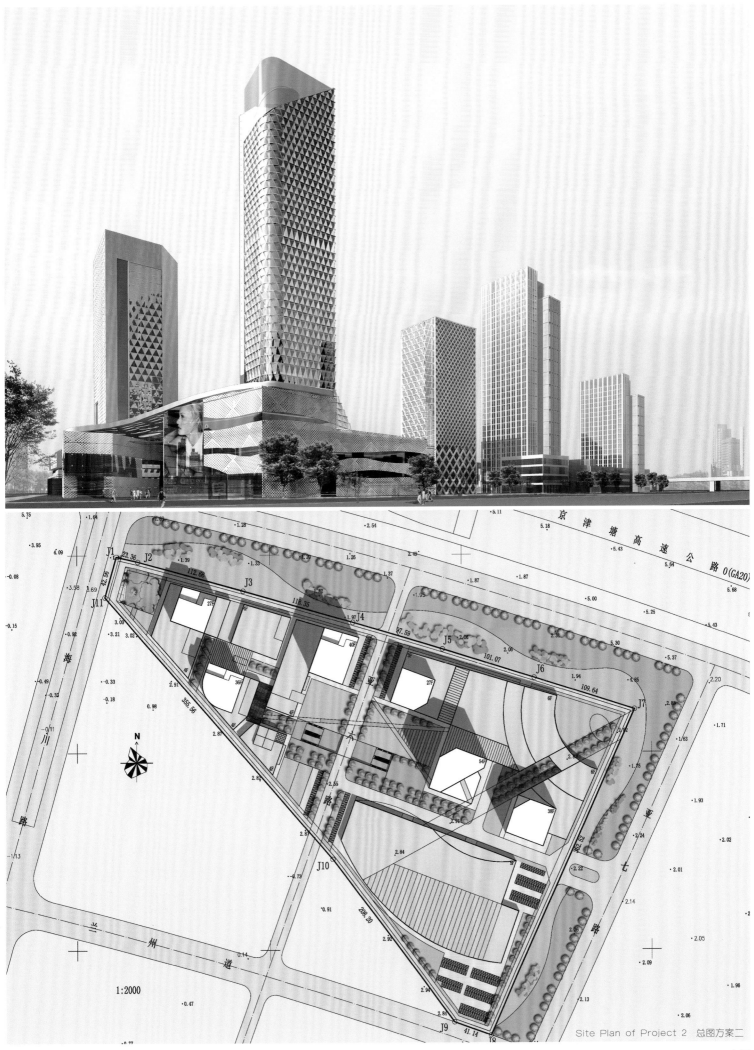

Site Plan of Project 2 总图方案二

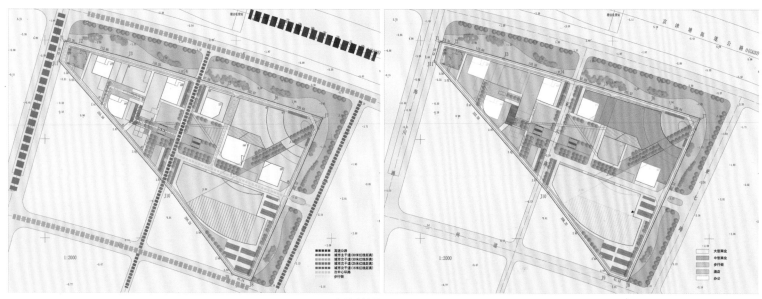

Traffic Analysis Drawing 交通分析图　　　　Functional Analysis Drawing 功能分析图

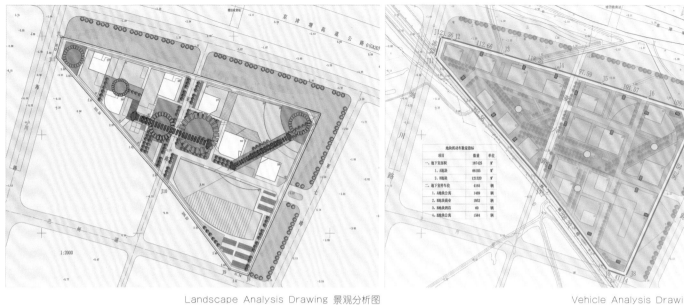

Landscape Analysis Drawing 景观分析图   Vehicle Analysis Drawing 机动车分析图

■ Architectural Design

Considering the triangular site, six high-rise buildings are arranged in two groups along the highway. At one corner of the site, large-scale commercial complex is set to form a central garden. Thus it has presented a high-quality and orderly living town. To comply with the development of modern economy and technology, the facades of the buildings are designed with a variety of materials, presenting varied textures as well as beautiful and tall buildings.

■ 建筑设计

方案巧妙地利用三角形的地界将6个高层建筑分为两大组团，沿高速公路一侧展开，余下的基地一角布置大型综合商业体，围合出内部的中空庭园，形成有序、高品质的都市群体风貌。为印证当代经济技术发展特征，建筑外立面采用多种材质，呈现出多种肌理，形成亮丽而又挺拔的建筑形态。

Taihe Urban Complex, Fuzhou
福州泰禾城市综合体

Keywords 关键词

Color and Proportion
色彩和比例

Material and Texture
材料和质感

Commercial Atmosphere
商业气氛

Location: Jin'an District, Fuzhou, China
Architectural Design: CCI Architecture Design & Consulting Co., Ltd.
Floor Area: 615,210 m^2

项目地点：中国福州市晋安区
建筑设计：上海新外建工程设计与顾问有限公司
建筑面积：615 210 m^2

Features 项目亮点

The whole development looks innovative and colorful with a strong sense of the times and advance. Simple and elegant style has shown the quality of an international commercial complex.

整体造型新颖、色彩鲜明，具有强烈的时代感和超前性，以简约、大气的风格体现出一个极具旗舰风范的国际商业建筑。

■ Overview

Occupying a total land area of 151,078 m^2, with Huagong Road on the south side, Lianjiang Road on the west side, phase IX of Helin New Town on the north side, the site will be used for commerce, business, hotel and public green.

■ 项目概况

基地位于福州晋安区化工路北侧、连江路东侧、鹤林新区九期南侧，总占地面积为151 078 m^2，用地性质为商业、商务办公、酒店用地和公共绿地。

■ Function Organization

The site is divided into three parts from east to west according to their different functions. Enjoying convenient traffic and big commercial interface, zone 1 is suitable for self-maintaining properties, including department store, complex building, entertainment building, internal commercial street, and two SOHO buildings. Zone 2 is the core of the site. Together with zone 1, it has formed a sales-type commercial block which includes retails, anchors, commercial plaza and three SOHO buildings. Zone 3 is in the west of the site. It is designed to be an relatively independent and quiet place which features five-star hotel and grade-5A office building.

■ 功能分区

根据业态类型，基地由东至西可大致划分为三个区：I区——交通条件最优，商业展示面最大，适合自持型物业，包含大型百货、综合楼、娱乐楼、商业内街、2栋SOHO办公楼；II区——基地核心，与I区商业结合设计销售型商业街区，包括零售店铺、次主力店、商业广场、3栋SOHO办公楼；III区——基地西段，设计为相对独立的静谧区域，主要包括五星级酒店与5A办公楼。

Aerial View 鸟瞰图

Site Plan 总平面图

Planning of Height Limit Sketch 规划限高示意图

Base Format Partition Analysis Diagram 基地业态分区分析图

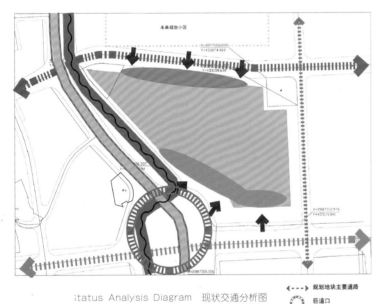

Status Analysis Diagram 现状交通分析图

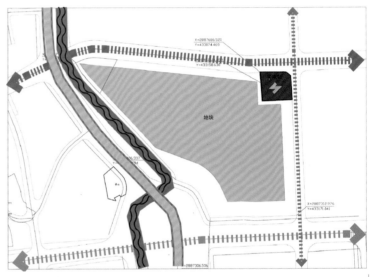

Base Status Analysis Diagram 基地现状分析图

Sale-oriented Business Layout 销售型商业布局图

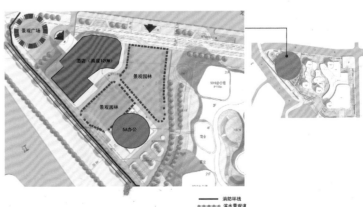

Hotel & Office 5A Layout 酒店及5A办公布局图

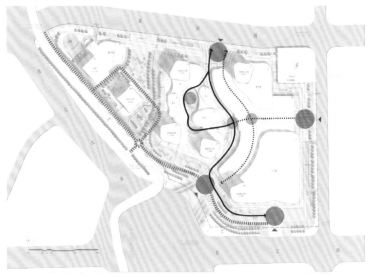
Pedestrian Flow Analysis Chart 人行流线分析图

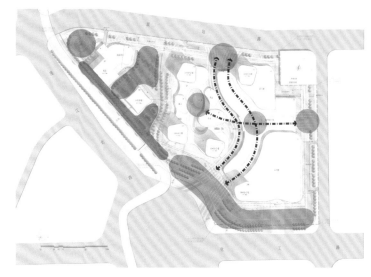
Analysis of Public Space Structure 公共空间结构分析图

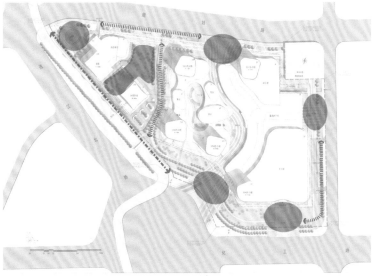
Analysis of Landscape Structure 景观结构分析图

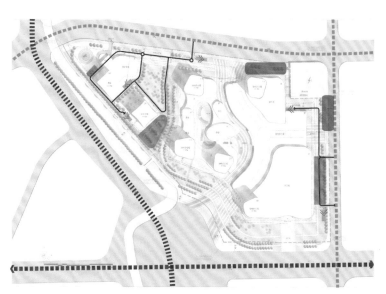
Vehicle Traffic Analysis Diagram 车行交通分析图

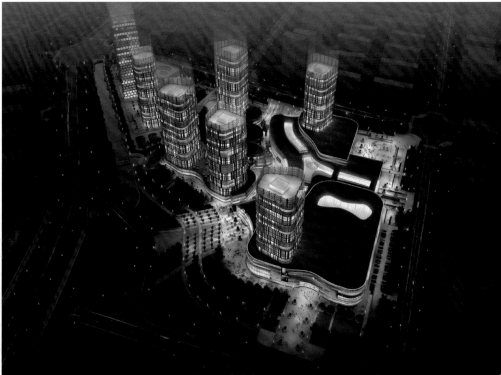

■ Architectural Design

The scale, color, material and texture of the buildings are carefully considered to present a beautiful outer appearance, highlight the personality and commercial atmosphere, and attract customers. The whole project looks innovative and colorful with a strong sense of the times and advance. Simple and elegant style has shown the quality of an international commercial complex.

■ 建筑设计

外观设计上通过对建筑的比例、色彩、材料和质感的综合运用，创造出美的形象，突出商业建筑的个性和商业气氛，吸引顾客。整体造型新颖、色彩缤纷，具有强烈的时代感和超前性，以简约、大气的风格体现出一个极具旗舰风范的国际商业综合体。

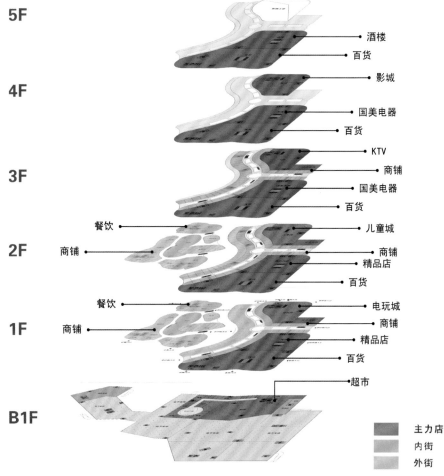

Commercial Format Partition Vertical Layout 商业业态竖向布局图

Basement Two Floor Plan 地下二层平面图

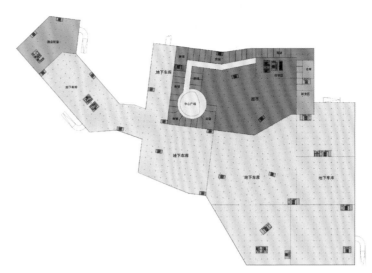
Basement One Floor Plan 地下一层平面图

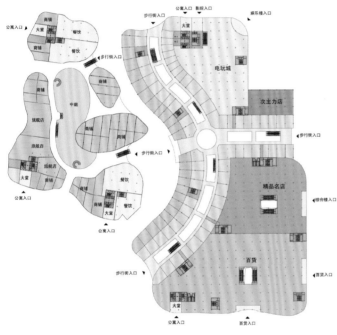
First Floor Plan 一层平面图

Second Floor Plan 二层平面图

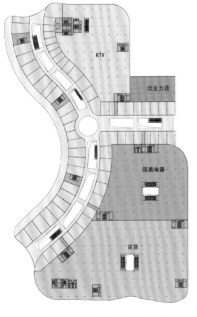
Third Floor Plan 三层平面图

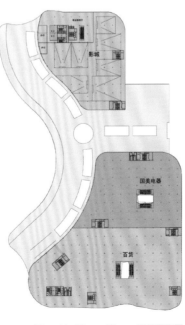
Fourth Floor Plan 四层平面图

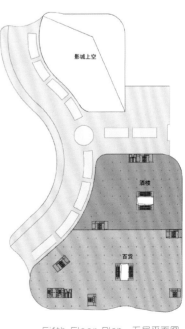
Fifth Floor Plan 五层平面图

Elevation 1 立面图 1

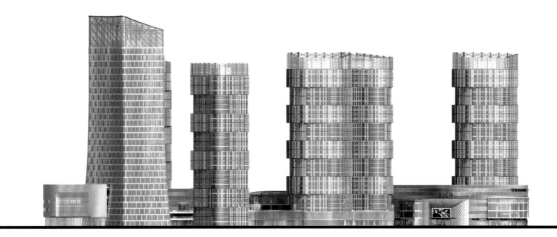

Elevation 2 立面图 2

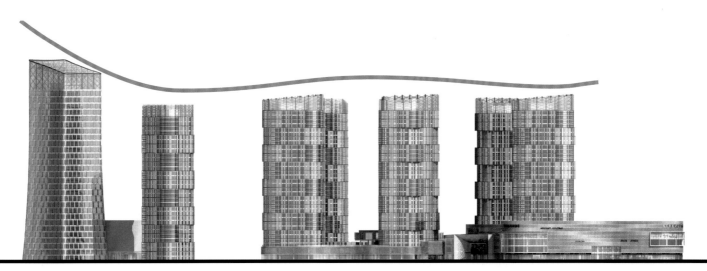

City Skyline Analysis Diagram
城市天际线分析图

Mojiazhuang Commercial Plot, Wuxi
无锡莫家庄商业地块

Keywords 关键词

Enclosed Layout
围合布局

Dynamic Space
活力空间

Commercial Atmosphere
商业氛围

Features 项目亮点

The design targets the young consumer group to create a more fashionable and vigorous shopping environment which also features high taste and remarkable appearance.

以针对年轻消费群体的策略实施设计，旨在打造一处更具活力和时尚度，同时兼具品质与内涵的地标性商业中心。

Location: Wuxi, Jiangsu, China
Architectural Design: CCI Architecture Design & Consulting Co., Ltd.
Floor Area: 185,789 m²

项目地点：中国江苏省无锡市
建筑设计：上海新外建工程设计与顾问有限公司
建筑面积：185 789 m²

■ **Overview**

Located in GuangYi area of Chong'an District, Wuxi, Jiangsu, the development is on the east of Jianghai Road and on the south of Fubei Road with a land area of 48,062 m². The site is divided into three parts, of which plot A occupies a land area of 17,468.9 m², plot B 18,349.2 m², and plot C 18,349.2 m².

■ 项目概况

本项目位于江苏省无锡市崇安区广益片区，江海路快速路以东，府北路以南，可建设用地面积约为48 062 m²。项目分为A、B、C三个地块，其中A地块用地面积约为17 468.9 m²，B地块用地面积约为18 349.2 m²，C地块用地面积约为12 244.5 m²。

■ **Overall Design**

According to the overall planning, the buildings are arranged in the periphery, with the central space left for a square which is also the main entry to plot A and C. The enclosed space creates a commercial atmosphere and increases the commercial value of these three plots. The spaces inside the buildings are flexible to cater to the youth's requirement for a fashionable and vigorous shopping environment. With "low carbon and energy saving" as the design principle, the interior of the commercial buildings makes full use of natural lighting and wind, and reduce the energy consumption of central air-conditioning and lighting, to be a low-carbon and energy-saving project.

■ 总体设计

在总体布局中，建筑沿外围排布，中心空间打开形成主要的人流集散广场，也是A、C地块的商业主入口。围合内部空间形成围合感，营造商业氛围，带动三地块整体的商业价值。建筑主体空间灵动，迎合年轻群体时尚、活力的环境消费心理。项目以"低碳节能"作为总体设计原则，在商业建筑内部尽量利用自然光照、通风，减少中央空调、照明的能耗，实现低碳节能的标志建筑。

■ Planning

The overall planning for plot A and B: in the northeast corner, a 27-storey office building is set to be the landmark in this area which guides people and encloses the internal space. Inside the two plots, there is internal street, courtyard-type headquarters or flagship stores. The first and second floor are used for exhibition, and the third floor is for offices, providing a comprehensive platform for brand enterprises. It has shown the image of the enterprises and products, and improved the quality of the whole project. The main entrances are set on Guangxiang Road and the planning road to lead people into the project.

Plot C is near to Wuxi-Shanghai floor stores on the south, and they are connected by a corridor. It is envisioned to be a specialty store for building materials and master designers. Devoting to create the No.1 building material and home furnishing mall on Xihu Road, it is designed with the sense of an indoor block. Corridors are the main elements to connect different parts and guide the flow of people, enhancing the commercial atmosphere.

Site Plan 总平面图

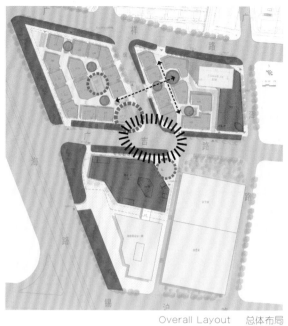
Overall Layout 总体布局

Function Analysis Diagram 功能分析图

Landscape Analysis Diagram 景观分析图

Fire Analysis Diagram 消防分析图

Overall Dynamic Traffic Analysis Diagram 总体动态交通分析图

Static Traffic Analysis Diagram 静态交通分析图

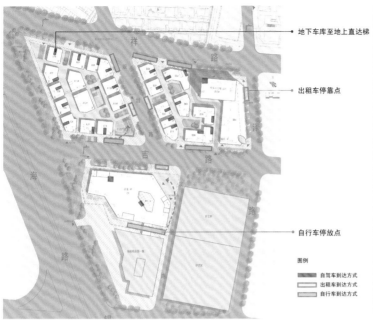
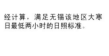

Arrival Mode Analysis 到达方式分析

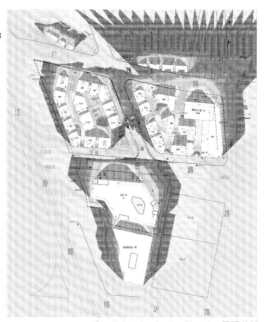

Sunlight Analysis Chart 日照分析

First Floor Plan of the Whole 整体一层平面图

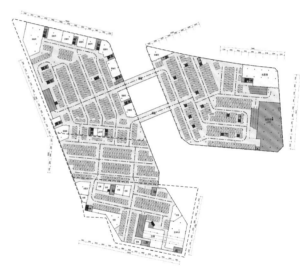

Basement Two Plan of the Whole 整体地下二层平面图

■ 规划布局

A、B地块整体规划布局：在东北角布置一栋27层的高层办公楼，在空间上打造地块标志建筑，承接外部人流，围合内部空间。两地块内部为内街、院落式的企业总部或品牌旗舰店，一、二层为展厅，三层为办公区域，为品牌企业提供综合平台。在打造企业及产品形象的同时，提高了整个项目的品质，主要出入口分别位于广祥路和规划路，形成统一的流线，相互导入人流。

C地块南临锡沪地板精品馆，通过连廊与地板精品馆联为一体，功能定位为建材家居馆及高品质设计师专门店，致力于在规模和品质上打造成为锡沪路第一大建材家居精品商城，在建筑平面上营造室内街区感。以廊为主题，从不同方向组织人流，形成回路，带动整个空间的商业氛围。

功能布局图

Basement Floor Plan of Land Parcel AC AC地块地下一层平面图

First Floor Plan of Land Parcel A A地块一层平面图

Sesond Floor Plan of Land Parcel A A地块二层平面图

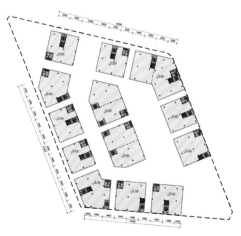

Third Floor Plan of Land Parcel A A地块三层平面图

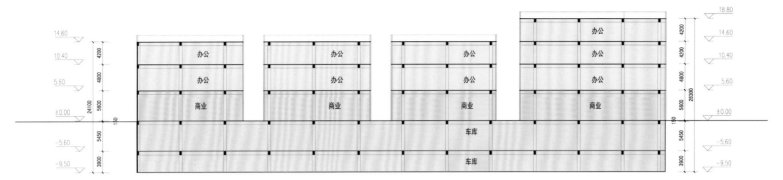

Section of Land Parcel A A 地块剖面图

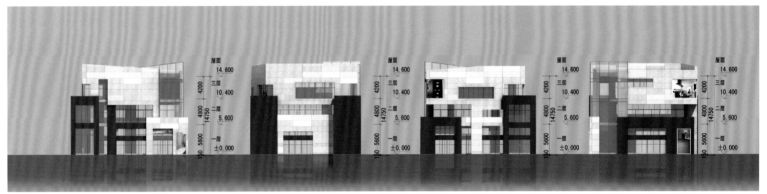

Elevation of A# Building A# 立面图

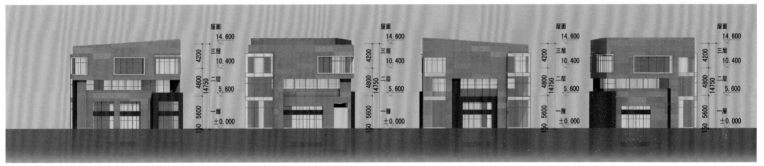

Elevation of A2/A7# Building A2/A7# 立面图

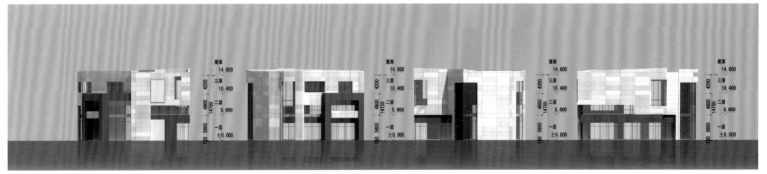

Elevation of A3/A6# Building A3/A6# 立面图

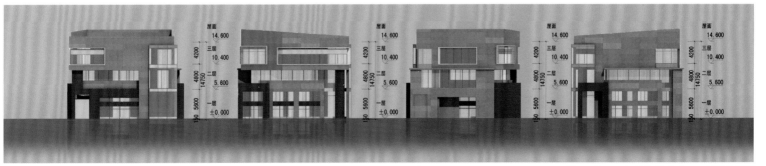

Elevation of A4# Building A4# 立面图

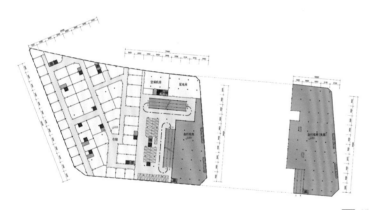

Basement Floor Plan of Land Parcel B　B地块地下一层平面图

First Floor Plan of Land Parcel B　B地块一层平面图

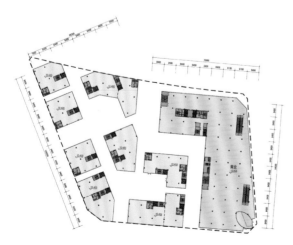

Third Floor Plan of Land Parcel B　B地块三层平面图

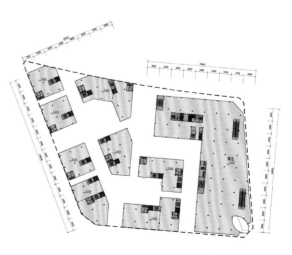

Second Floor Plan of Land Parcel B　B地块二层平面图

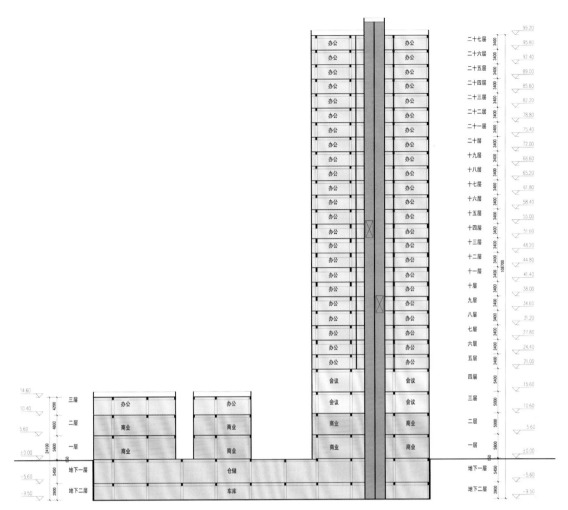

Section of Land Parcel B B 地块剖面图

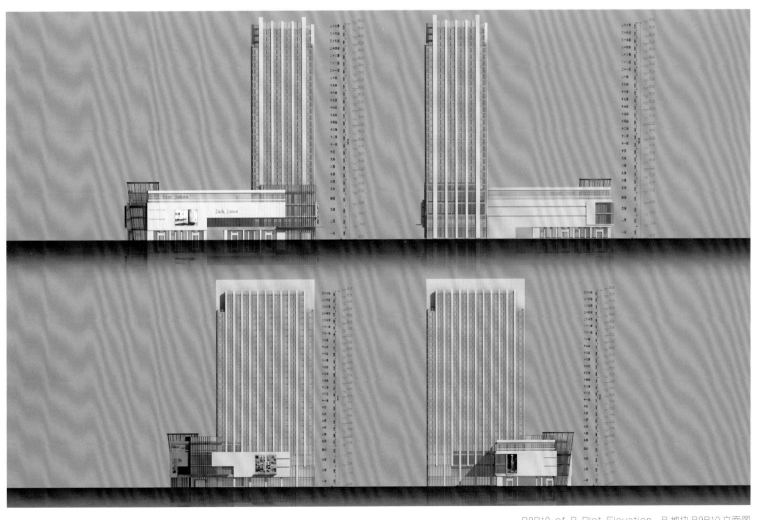

B9B10 of B Plot Elevation B 地块 B9B10 立面图

183

First Floor Plan of Land Parcel C　C地块一层平面图

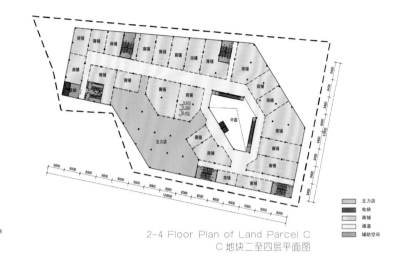

2-4 Floor Plan of Land Parcel C
C地块二至四层平面图

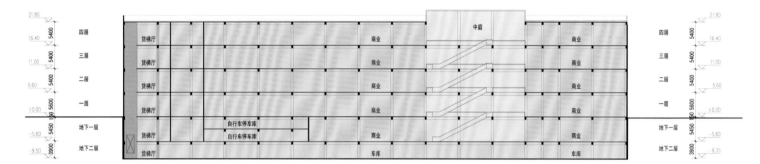

Section of Land Parcel C　C地块剖面图

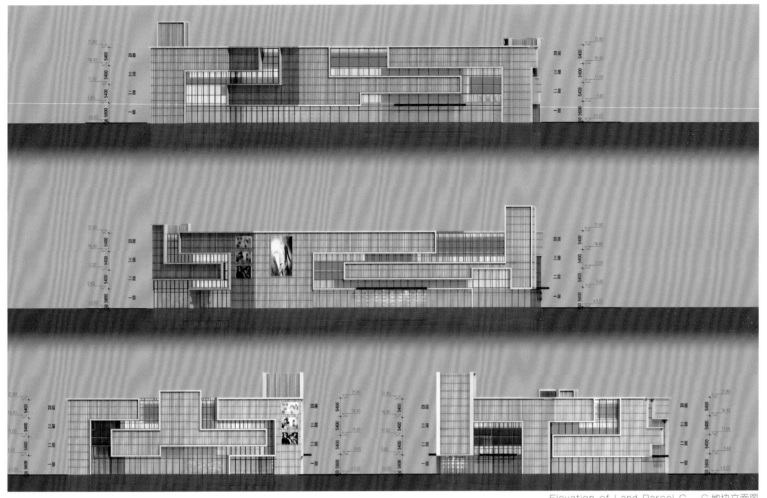

Elevation of Land Parcel C C地块立面图

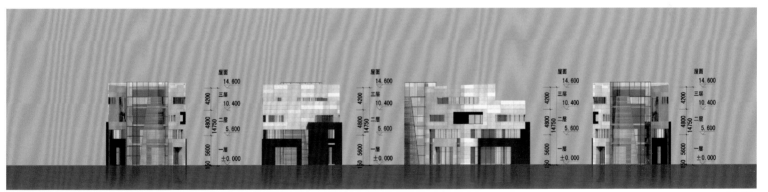

Elevation of B1# Building B1#立面图

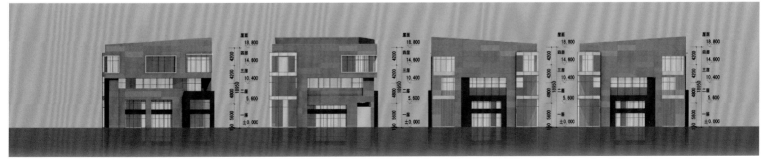

Elevation of B2# Building B2#立面图

Trans-surface, Beijing
北京"超表皮"

Keywords 关键词

Structural System
结构系统

Support System
支持系统

Interface System
界面系统

Features 项目亮点

Trans-surface pays close attention to create an open interface that can satisfy the needs of different clients. Trans-surface is comprised of three main systems, structural system, support system and interface system.

"超表皮"关注的重点是试图创造一个开放的界面能够适应不同用户的需求。"超表皮"是由三个主要的系统复合而成：结构系统、支持系统和界面系统。

Location: Beijing, China
Architectural Design: CU Office
Architects: Che Fei, Jin Hao, Feng Jun, Wang Hezhao, Alexander Nestle
Area: 11,600 m²

项目地点：中国北京市
建筑设计：超城建筑师事务所
设计人员：车飞、金颢、冯军、王禾昭、Alexander Nestle
面　　积：11 600 m²

■ **Overview**

Trans-surface pays close attention to create an open interface that can satisfy the needs of different clients. It builds an interactive window between virtual space and real space that offers possibility enabling user to shuttle back and forth, bringing about openness and delightful experience for urban public space again. It makes people to shift their attention from physical space to the interactive experience between network and hyperspace.

■ **项目概况**

"超表皮"关注的重点是试图创造一个开放的界面能够适应众多不同用户的需求。这个界面能够在虚拟空间与真实空间之间建立互动的窗口，提供使用者穿梭于其间的可能性，重新为城市公共空间带来广阔的开放性与令人愉快的体验。项目使人们从对城市中物理边界的关注转向对网络与多维空间的交互体验。

■ **System Design**

Trans-surface is comprised of three main systems, structural system, support system and interface system. It blurs the boundary of virtual world and real world, makes the daily life and information technology closely linked still further and enlarges the possibility of information Web service, enabling virtual world influence and even change the physical world.

The project unites information design with urban public space. And Trans-surface becomes the unique urban life generator, which responses the theme of this competition, let pleasure back to the block and let the block back to life.

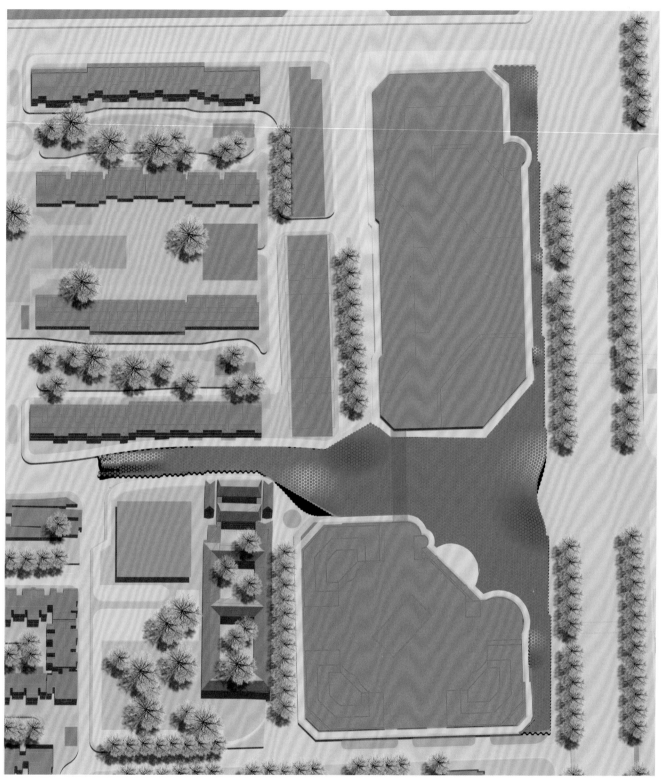

Site Plan 总平面图

■ 系统设计

"超表皮"是由三个主要的系统复合而成,它们是结构系统、支持系统和界面系统。"超表皮"将真实世界与虚拟世界的边界模糊化,使人们的生活与信息技术的联系更为紧密,扩大了信息网络服务的可能性,使得虚拟世界可以影响甚至改变物理世界。

项目将信息设计与城市公共空间相结合,"超表皮"成为了独特的"城市生活发生器",并以此回应此次竞赛的主旨:"让乐趣回归街区,让街区回归生活。"

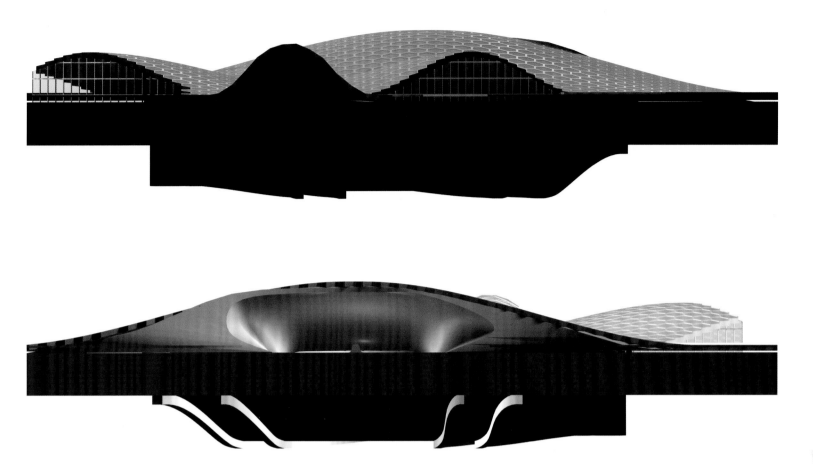
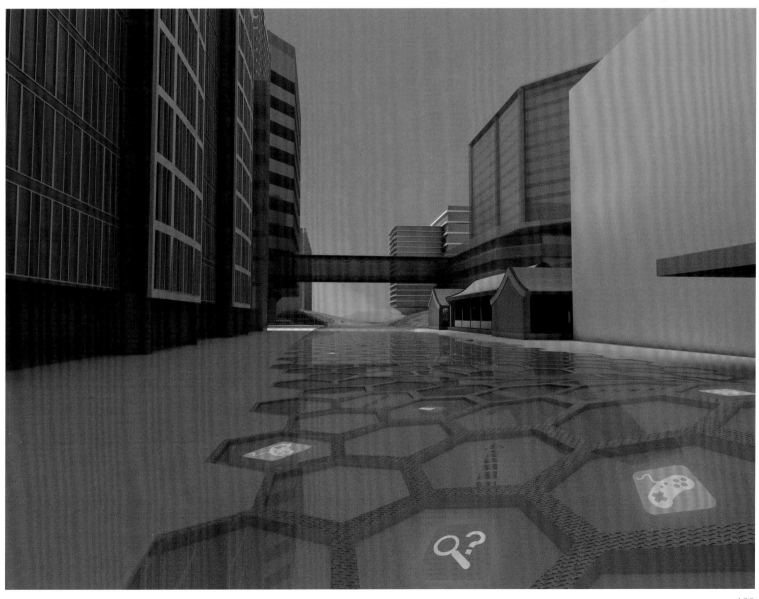

Delta Area of Jintang County, Sichuan
四川金堂县三角洲、三星片区

Keywords 关键词

Eco Idea
生态理念

Enclosed Building
围合式建筑

Cultural Connotation
文化内涵

Features 项目亮点

It aims to continue the road context and pleasant proportion of the old town. Local traditional enclosed architectural style and form are also kept to enhance the cultural connotation and the feeling of belongings.

在设计中，主要考虑延续老城区的道路肌理和宜人的空间尺度，并适当地延续当地典型的围合式建筑风格和建筑形式，以此增加城市的文化底蕴和市民的归属感。

Location: Jintang County, Sichuan, China
Planning: United Design Group
Project Leader: Wang Weiliang
Architects: Ma Teng, Xiao Fan
Land Area: 16,404,100 m²

项目地点：中国四川省成都市
规划设计：UDG联创国际
项目负责人：王卫良
规划设计人员：马腾 肖凡
占地面积：16 404 100 m²

■ Overview

The design pays attention to the integration of urban spaces. In addition to the administrative center and commercial center, and on the occasion of the development of Delta area it has complemented the city with a public center which integrates business, finance and convention and exhibition.

■ 项目概况

本项目在设计中，注重对城市空间结构的整合，在行政中心和商业中心的基础上，以三角洲、三星片区的开发为契机，增加一个集商务、金融、会展于一体的综合性城市公共中心。

■ Design Concept

The planning is based on eco ideas. Green corridors, city park and low-density development zone are designed to establish connection with Longquan Mountain, Beihe River, Zhonghe River and other eco parks in the city. And between Beihe river and Longquan Mountain, more eco corridors are added to integrate the city with mountain and water.

It aims to continue the road context and pleasant proportion of the old town. Local traditional enclosed architectural style and form are also kept to enhance the cultural connotation and the feeling of belongings.

■ 设计理念

同时，规划基于生态理念。通过绿化廊道、城市公园、低密度开发区等设施，打通龙泉山脉和北河、中河以及城市内生态公园的联系。在北河和龙泉山脉之间，更进一步增加多条生态廊道，最终实现山、水、城的有机融合。

在设计中，主要考虑延续老城区的道路肌理和宜人的空间尺度，并适当地延续当地典型的围合式建筑风格和建筑形式，以此增加城市的文化底蕴和市民的归属感。

Aerial View 鸟瞰图

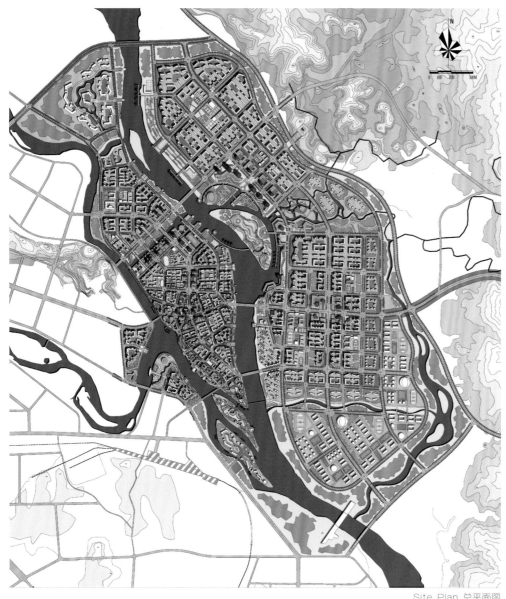

Site Plan 总平面图

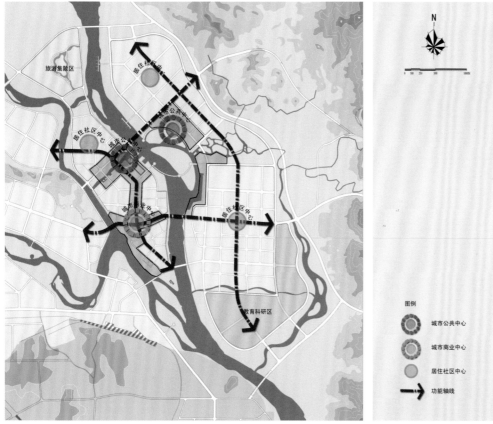

Structure Planning Drawing 规划结构图

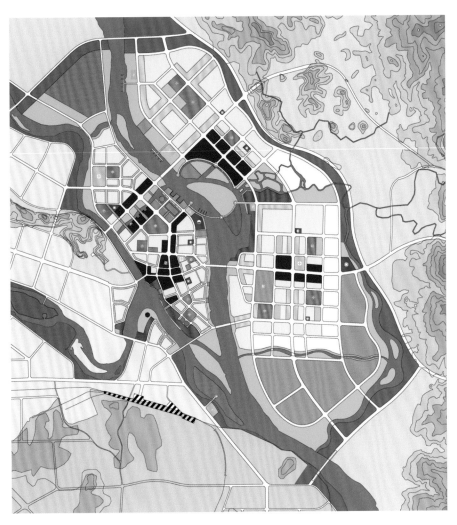
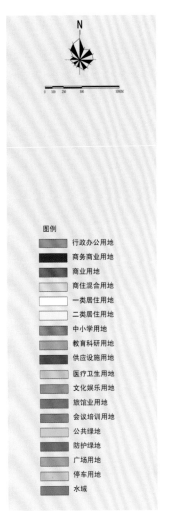

Land Use Diagram 土地利用图

Road System Diagram 道路系统图

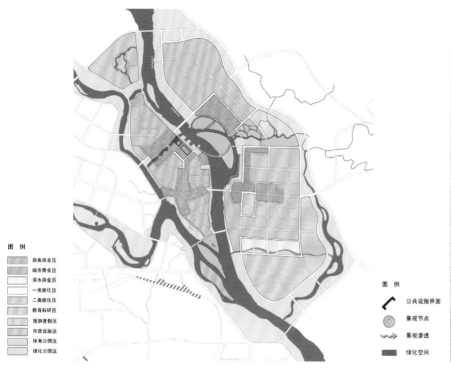

Functional Space Analysis Drawing
功能分区图

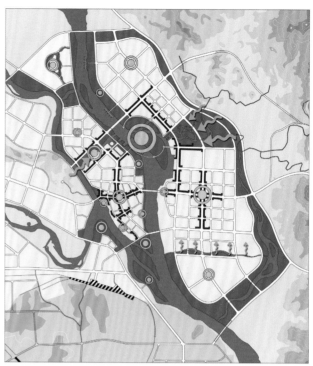

Open Space System Drawing
开放空间系统图

Minmetals International, Tianjin
天津旷世国际

Keywords 关键词

Twin Tower Modeling
双塔造型

Arc Interface
弧形界面

Exquisite Details
细部精美

Features 项目亮点

Surface treatment of the towers pursues to be concise and lively. Thrown elements produced new levels and dynamic that makes the volume huge but interesting, majestic, confident and pleasant.

大厦的表皮处理追求简洁明快的基调，外立面错动的元素产生新的层次感和活跃感，使得建筑体量虽大却饶有趣味，挺拔有力且细部精美。

Location: Minmetals Property, Tianjin, China
Architectural Design: ZPLUS
Architect: Liu Shunxiao
Total Floor Area: 183,800 m²

项目地点：中国天津市滨海新区
建筑设计：ZPLUS普瑞思建筑规划设计公司
主设计师：刘顺校
总建筑面积：183 800 m²

■ **Overview**

Located in Tianjin Xiangluowan Business District, the project is at the intersection of south-north central axis in Binhai New Area and east-west high-tech industrial zone. The project is a twin tower over 120 m high. Tower A is a high-end office building, Tower B is service apartment. Total floor area is 183,800 m², among which, 145,000 m² is aboveground part, while 38,500 m² is underground part.

■ **项目概况**

项目位于天津响锣湾商务区，处在天津滨海新区商贸南北向中间轴和高新技术东西向产业带的交汇点。五矿竞标项目呈现双塔造型，两个主塔楼高度均在120 m。A座为高端写字楼，B座为酒店式公寓，总建筑面积为183 800 m²，其中地上建筑面积约为145 000 m²，地下面积为38 500 m²。

■ **Planning Layout**

The 30-storey apartment building is in the south of the site in shape T. Located at the cross, its spectacular image becomes the symbolic urban landscape. While in the premise of ensuring enough area, it provides more southward houses and indoor exchange space such as hanging garden. The other 28-storey building in the north is a complex with apartment and offices, which makes the best use of the terrain of the site. Internal link is produced with the same curve composition of the two towers. Semi-enclosed space and unified urban interface shape unified and changing urban interface landscape, highlighting the optimistic spirit of creation of the twin tower.

■ **规划布局**

在规划上面，公寓楼共有30层，采用T字形平面布置在基地南侧，在主干道十字交口处产生一高耸壮观的形象，成为地标性的城市景观，同时在保证面积足够的前提下，提供更多南向户型和室内交流空间，如空中花园。另一高层采用一字形略带弯曲的平面布置在基地北侧，为公寓及办公楼综合体，共28层。此布局能最大限度地利用基地地形，同时两个楼体以相同的曲线构图产生内在联系，在街区内侧形成一气呵成的弧形半围合空间和统一的都市界面，在基地沿主干道一侧，形成既统一、又多变的景观，烘托出双塔建筑不拘一格、乐观向上的创意精神。

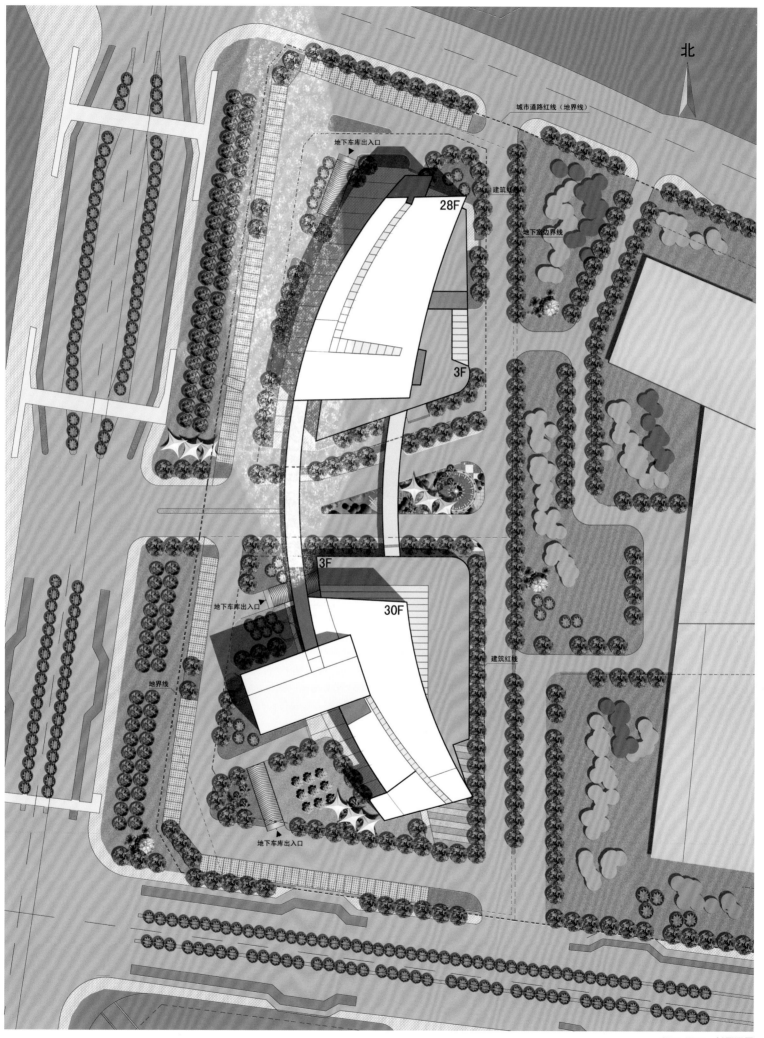

Site Plan 总平面图

Structure Planning Drawing 结构分析图

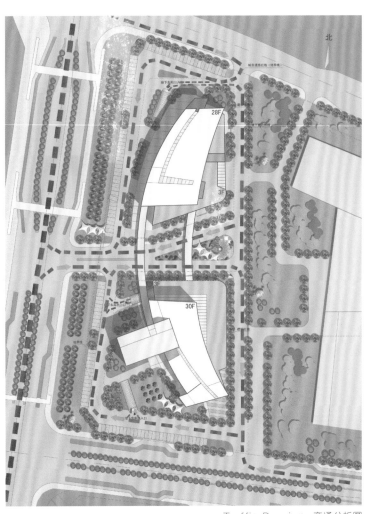

Traffic Drawing 交通分析图

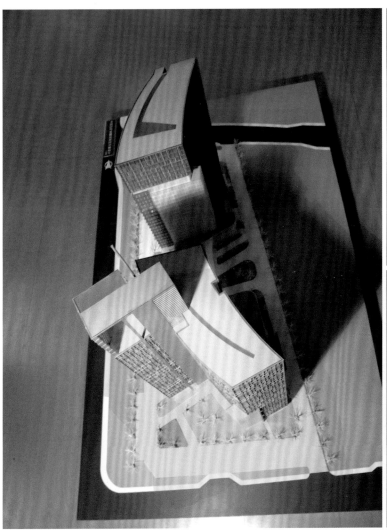

First Floor Plan 首层平面图

■ Architectural Design

In terms of architectural design, two towers formed an integrated arc-shaped interface and distinctive group comes into being with the help of intensive dynamic in local part. The shape T gets rid of the tedium of pure plate-type high-rise and makes the arc-shaped group much more vivid and versatile. Arc shape not only acts in cooperation with the terrain, but brings the energetic spirit that makes the building dynamic and amiable.

Surface treatment of the towers pursues to be concise and lively. Abstract geometric shape, hanging garden and skin texture formed various comparisons, which leave you a deep impression. Thrown elements produced new levels and dynamic that makes the volume huge but interesting, majestic, confident and pleasant.

In terms of design process, three-dimensional consciousness is taken root in. Various motive power, sliding movement, transformation and tracks make the building magic box. The comparison of straight face and curve face enriched the visual and spatial perception.

■ 建筑设计

在建筑设计上面，两座塔楼形成一个完整的弧形界面，局部有高塔增强了力度，形成了具有鲜明个性的组群。尤其南部大楼的T字形平面产生出的富于变化的体量，使高层充满挺拔向上的力度感，一扫单纯板式高层的单调，使弧形的群体更加生动，富于变化。弧形不仅呼应了地形特点，也带来如乘风破浪的昂扬精神，并使得大体量的建筑充满了动态和亲和感，整个群体简洁且易于识别弧线、塔楼、天桥，配合高层特有的自然升腾感与力度。

大厦的表皮处理追求简洁明快的基调，抽象的几何形状，空中花园与粗细对比的表皮质感形成多种对比，令人印象深刻，同时暗示了矿石的斑驳肌理，与公司的主业产生一些联系。错动的元素产生新的层次感和活跃感，使得建筑体量虽大却饶有趣味，挺拔有力且细部精美，雄伟、自信、令人愉悦。

在设计的过程中，设计师将三维意识深深地根植于此设计中，多种原动力、滑动、变形、轨迹将大厦塑造成具有魔力的盒子，直面和曲面的相互对比导致了作品中视觉和空间感受的丰富性，能量、冲击感在高耸的塔楼上、在曲线和直线的并置中得以呈现。

201

Elevation 1 立面图 1

Elevation 2 立面图 2

Wanlong Shenyang (Shenbei) Project
沈阳万隆（沈北）项目概念设计

Keywords 关键词

Neoclassicism
新古典主义

Modern Materials
现代材料

Traditional Style
传统样式

Features 项目亮点

Modern techniques and materials are used to create buildings with traditional-style outlines. It also pays attention to decoration effect and uses classical elements to create a special atmosphere.

项目用现代材料和技术手法追求传统式样的大致轮廓。同时还注重装饰的效果，运用古典装饰元素和设施来烘托环境气氛。

Location: Liaoning, Shenyang, China
Architectural Design: Agence C&P Architecture
Architects: Lan Jian, Li Mengyao
Total Land Area: 989,000 m²
Total Floor Area: 2,383,200 m²

项目地点：中国辽宁省沈阳市
建筑设计：C&P（喜邦）国际建筑设计公司
设计人员：兰剑　李梦瑶
总占地面积：989 000 m²
总建筑面积：2 383 200 m²

■ **Overview**

The project is located in Shenbei New Area of Shenyang City. Situated in Hushitai District of Puhe New Town, at the intersection of Yalujiang North Street and Shenbei Road, it takes only 30 minutes to drive to the downtown. Shenbei Road leads to Daoyi Development Zone and Huishan Development Zone on the south; Yalujiang North Street leads to the downtown on the east. The completion of Shenyang North Station and the metro line will bring more convenient transportation. The whole project consists of international residential community, commercial city, hotel, cultural exhibition center, etc.

■ 项目概况

项目位于沈阳城北，沈北新区内。项目处于沈北新区蒲河新城的虎石台片区，骨架路鸭绿江北街与沈北路的交界处，距离市中心约30分钟的车程，南侧沈北路直达道义和辉山开发区，东侧鸭绿江北街直接市中心，未来沈阳北站以及地铁线路建成后将使得交通更加便捷。整个项目包括国际社区、商业城、酒店以及文化展览中心等。

■ **Planning**

The design has followed the rules of new urbanism to advocate traditional neighborhood model. Varied public spaces and facilities are set within a short distance from the public stations. This walkable space design will greatly reduce relying on cars. The cluster center is only 5-minute-walk to the borders. The building clusters are set along the central landscape corridor, and public buildings and commercial facilities are arranged in the central area which is connected with the commercial street on the south side. The clusters are relatively independent with separate functional systems. They will be connected together in the future development.

■ 规划布局

本案在设计中依循了新城市主义中所倡导的传统邻里结构，将各种公共活动空间和公共设施布局于公共站点的步行距离之内，通过对适宜步行的空间设计，减少对于汽车的依赖。其组团的最佳阵容是中心到边界直线距离为5分钟的路程。各组团在中心景观（生态长廊）的两边，呈梯状向两边分布。在中心区域范围之内布置居住区级的公建和商业配套设施，与南侧的商业街形成连续商业带。每个组团各自独立，其功能各成体系，在后期开发没形成之前可自行运营，最终又会在后期的发展中相互联系，互补完整。

Aerial View 鸟瞰图

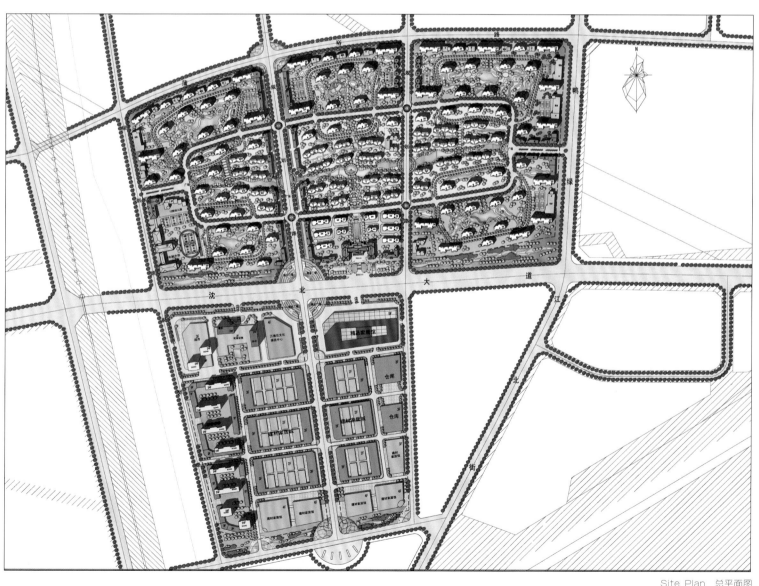
Site Plan 总平面图

Plot Peripheral Traffic Analysis　地块周边交通分析

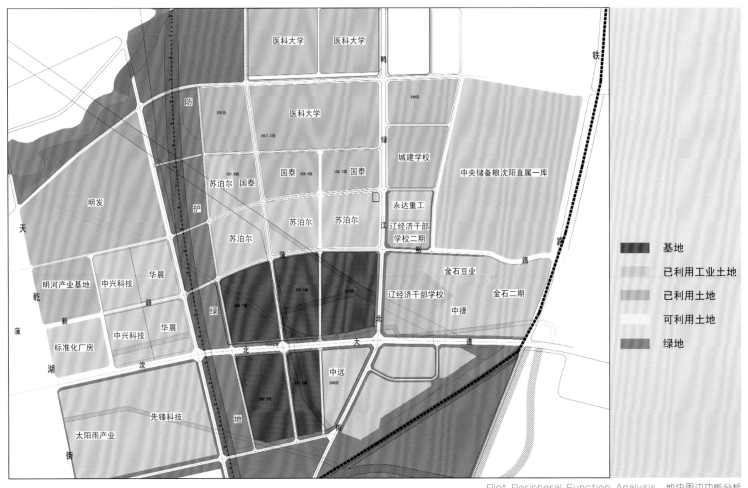

Plot Peripheral Function Analysis　地块周边功能分析

Phase Sketch 分期示意

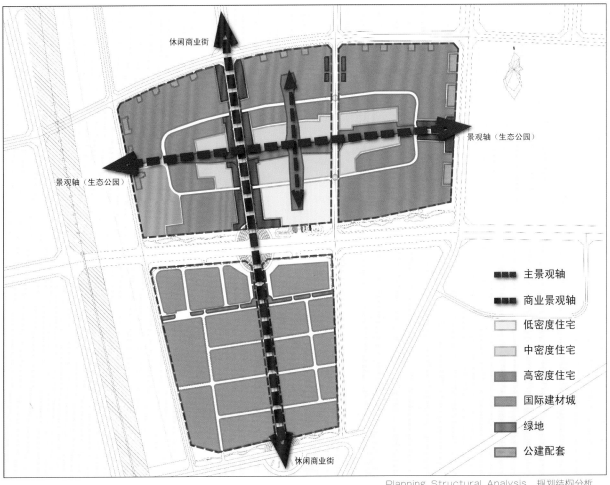

Planning Structural Analysis 规划结构分析

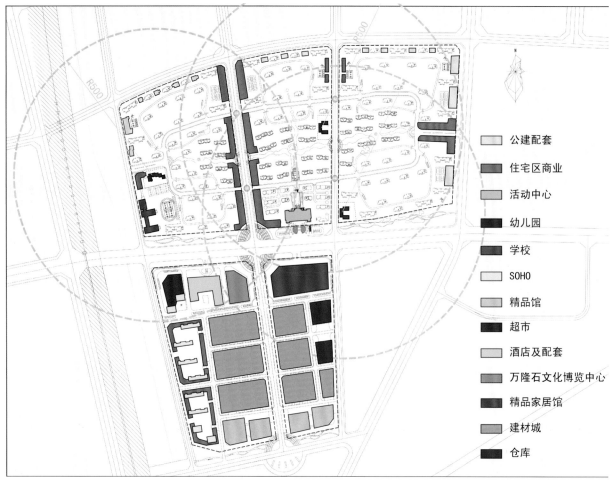

Public Building Layout 公建分布

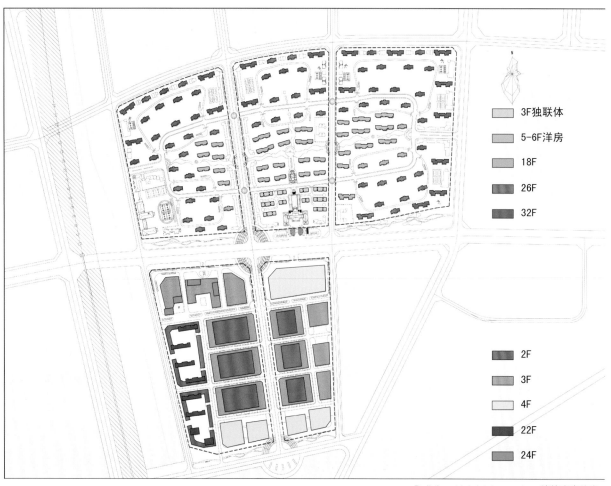

Building Height Layout 建筑高度分布

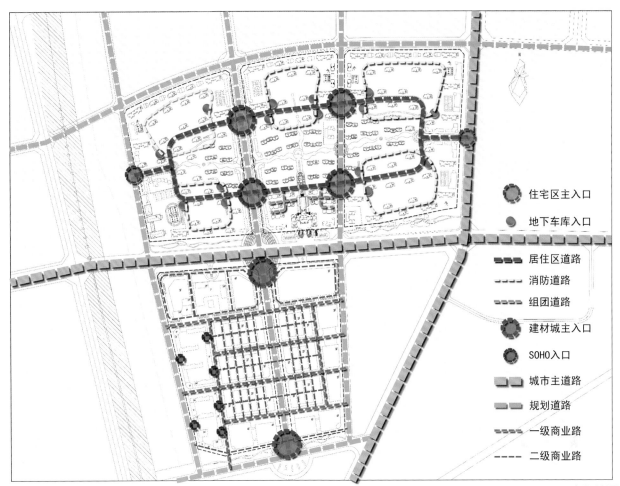

Traffic Drawing 交通分析图

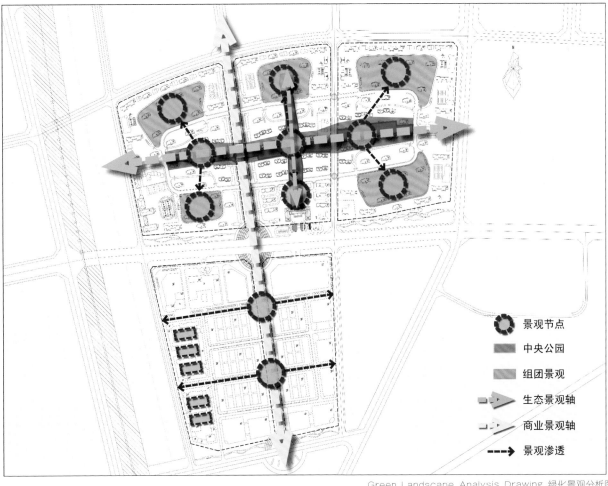

Green Landscape Analysis Drawing 绿化景观分析图

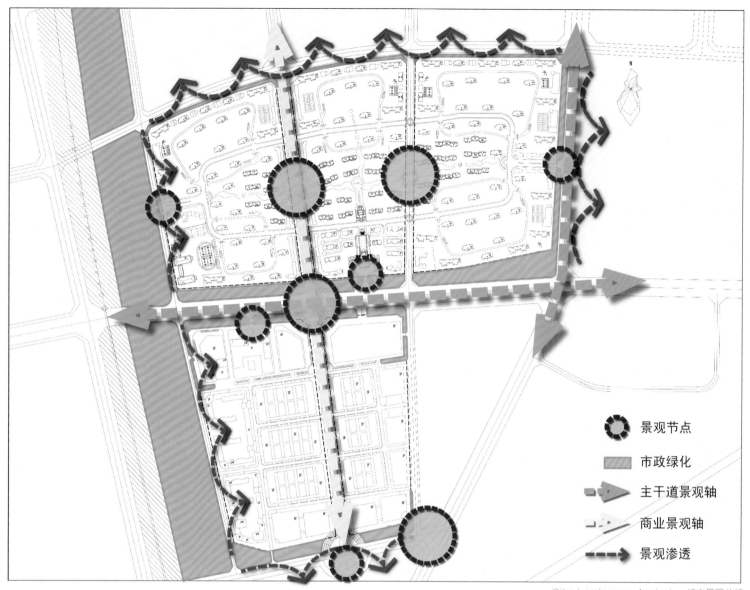

City Landscape Analysis 城市景观分析

虽然周边学校较多，但满足服务半径的学校没有，均在2000米以外，路途遥远，且现状周边多为村镇学校幼儿园，缺少高品质名校。

教育设施
- 中学
- 小学
- 幼儿园

其他配套服务
- 虎石台基督教堂
- 益民医院
- 银行
- 超市
- 派出所
- 邮政储蓄
- 沈煤宾馆

周边最近的服务配套均在虎石台镇，距离也在2000米以上，没有现成可利用配套设施，比较不便。

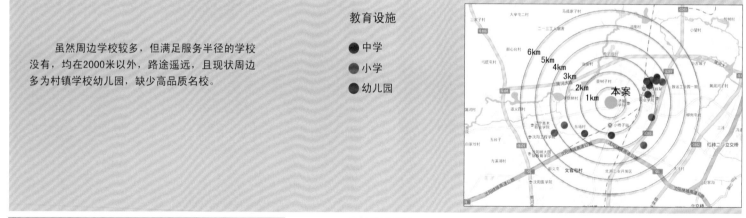

Comprehensive Facilities Status Analysis 综合设施现状分析

■ Architectural Style

Designed in neo-classical style, it pays attention to complete artistic form and sculptural appearance, and tries to be elegant, magnificent and harmonious. The buildings emphasize the decoration effect, applying modern techniques and materials to create elegant ambiance. In addition, simple techniques, modern materials and artistic skills are adopted to form a traditional outline. Classical decoration elements and facilities are used to create classical atmosphere.

■ 项目风格

项目采用新古典主义的风格，注重古典艺术形式的完整、雕刻般的造型，追求典雅、庄重、和谐。整个建筑注重装饰的效果，用现代的手法和材质还原典雅气质，使其具备古典与现代的双重审美享受。建筑造型追求神似，用简洁的手法、现代材料和艺术技术手法追求传统式样的大致轮廓的特点。同时还注重装饰的效果，运用古典装饰元素和设施来烘托环境气氛。

■ Landscape Design

In terms of landscape design, the east-west eco corridor is the highlight of this project which enables the residents to feel park-like environment at the door. There are two levels of green systems. Level one is the central landscape, and level two is the green spaces inside the building clusters. Additionally, the green belts along the roads have connected all the landscapes together.

■ 景观设计

在绿化景观方面，贯穿东西的生态长廊是本案的一大亮点，出门即可感受到公园，让住在这里的居民充分感受到自然。项目景观采用两级绿化体系，一级为居住区级的中心绿化，二级为组团内部的绿化空间，而居住区级道路两侧旁边设置的扩大的绿化带，则起到了联系整个景观的作用。

Side Street South Elevation 南立面沿街图

Side Street North Elevation 北立面沿街图

Side Street East Elevation 东立面沿街图

Side Street West Elevation 西立面沿街图

Poly Real Estate's Yudong District Project Design
保力地产御东区项目规划设计

Keywords 关键词

Grid Construction
网格系统

Vision Dislocation
视觉错位

Massive Building Massing
体量巨大

Features 项目亮点

On the view of facade design, the architectures regard the building as a whole by using the grid construction to create a vision dislocation on the facade, also cut down the building massing to a great extent.

在立面设计上，设计师将建筑作为一个整体来考虑，运用了一套网格系统，使立面呈现视觉错位，很大程度上消减了建筑的巨型体量感。

Location: Datong, Shanxi, China
Architectural Design: Graft Beijing Co., Ltd

项目地点：中国山西省大同市
规划设计：北京格拉福特建筑咨询有限公司

■ **Overview**

This project will have a land area of 370,000 m², and the floor area is 80,000 m², with 4,000 apartments, 40,000 m² of office space, 2 hotel buildings, over 60,000 m² for the commercial space, a kindergarten and an eight-class primary school.

■ **项目概况**

项目要在占地面积为370 000 m²、建筑面积约为80 000 m²的规模下发展超过4 000套公寓、40 000 m²的办公空间、两栋酒店建筑、超过60 000 m²的商业零售空间以及一个幼儿园和一个八班制小学。

■ **Planning**

Not the same as the traditional single tower, the architectures put all the buildings around the base. Making all the buildings to be unique and special cuboids and surround the base, the architectures want to avoid any negative atmosphere will be caused by a single tower, and make the interiors of the base more open and create a dynamic space with a sense of surrounding. This kind of concept, which using some cuboids buildings to create a wide natural landscape in the center of them, is called "Reverse Walls".

■ **规划布局**

不同于传统的单塔，设计师将所有的建筑沿基地周边布置。通过这种将所有的建筑体块压缩成极端而独特的长条形，并且沿基地边缘布置的方式，避免了单塔之间容易出现逸散消极空间的同时，解放了整个基地内部，并创造一种围合感强烈的积极空间。这种连续的长条形建筑围合中心广阔的自然景观的构想，谓为"反转的城墙"。

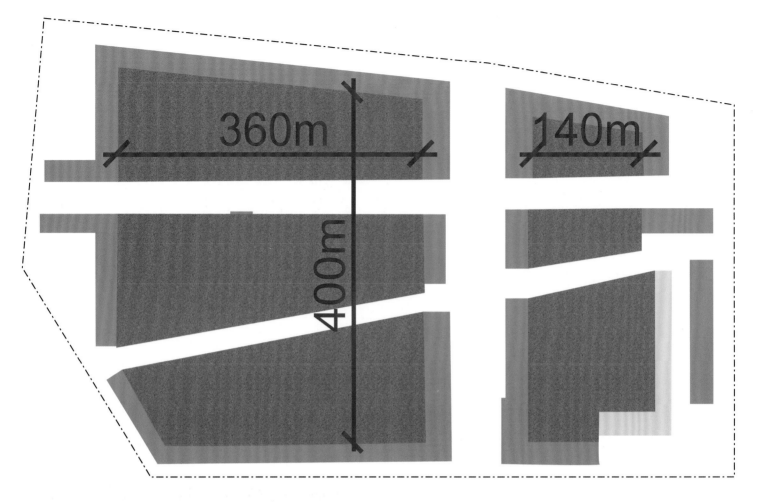

■ Facade Design

In the terms of facade design, the architectures regard the buildings as a whole and have create a grid construction. By having 9 units in a grid, also one unit with 9 grids, it will makes people confused whether the building is 9 times of the actual size or 1/9. This kind of vision dislocation have attracted people's interesting to the abstract game of the grid, and have weaken the massive building massing to a great extent.

The grid is made of the insulated rendered concrete on the facade of block. There have 9 simple and repeated grids on every single unit, and the grid is occupied by the windows or the Exterior Finish Insulation System (EFIS). With this kind of construction, it will be easy to cut down the window rate to 30% on the north, and 60%~70% on the south.

New York City Central Park 纽约中央公园

Beijing Tiananmen Square 北京天安门广场

■ 立面设计

在立面设计上，设计师将建筑作为一个整体来考虑。因此，设计师发展了一套网格系统，在这套系统下，有的地方9个单元组合到一个网格中，有的地方一个单元中就提供了9种网格。最后人们会分不清楚到底该建筑是实际尺寸的9倍还是1/9。这种视觉错位在很大程度上消减了建筑的巨型体量感，让人们的注意力转移到网格的抽象游戏中去。

网格是由附在主体立面上的保温抹灰混凝土结构构成。主立面是每个单元上简单重复的九宫格。这些九宫格被窗户或者保温抹灰实墙（EFIS）填充。该系统将会非常容易地降低北面开窗率至30%，而南边开窗率达到60%～70%。

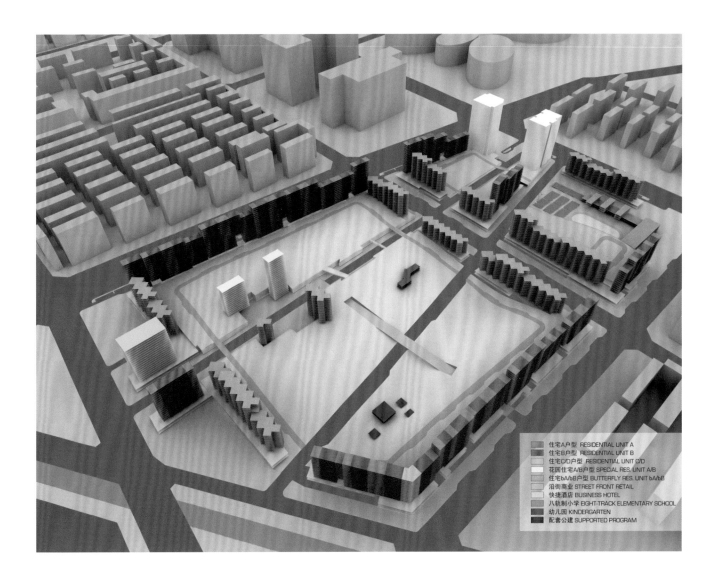

■	住宅A户型 RESIDENTIAL UNIT A
■	住宅B户型 RESIDENTIAL UNIT B
■	住宅C/D户型 RESIDENTIAL UNIT C/D
■	花园住宅A/B户型 SPECIAL RES. UNIT A/B
■	住宅bA/bB户型 BUTTERFLY RES. UNIT bA/bB
■	沿街商业 STREET FRONT RETAIL
■	快捷酒店 BUSINESS HOTEL
■	八轨制小学 EIGHT-TRACK ELEMENTARY SCHOOL
■	幼儿园 KINDERGARTEN
■	配套公建 SUPPORTED PROGRAM

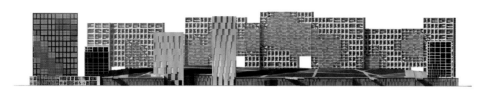

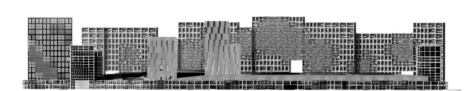

South Elevation 南立面图

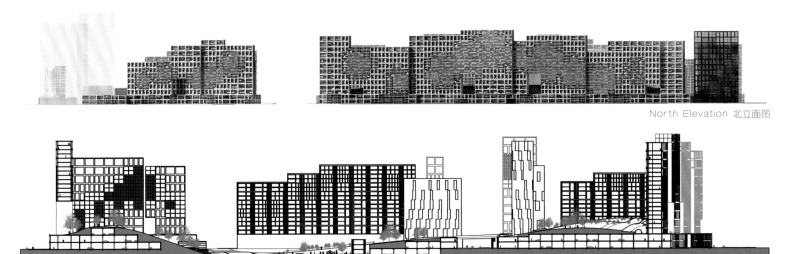

North Elevation 北立面图

Section 剖面图

■ House Design

There are 58% of apartments located on the north-south buildings. The buildings are with 8.4m grid and 2.9m height. The whole buildings are made of lots of 4 m x 8.4 m units, and every single units, and 4 apartments of 102 m² and 76 m² on every floor. Two of the apartments extend from south to north, to achieve natural ventilation. The smaller centre apartments have been equipped with cross ventilation ceiling, and pass through the corridor partly.

The rest 42% are located on the east-west buildings, but turned to south by 45 degree and in a zigzag pattern. In order to make every single apartment to face south in 45 degree, the architectures have made the depth to be 25 m when necessary, which can set a 5 m x 6 m light well and ventilation shaft in the center to provide cross ventilation for all the apartments.

■ 户型设计

58%的公寓位于南北向的建筑内。建筑的柱网为8.4 m，层高为2.9 m。整个建筑由多个4 m x 8.4 m 的单元组成，每个单位中包含4个公寓，面积分别为102 m²和76 m²。四个单元中的两个从南到北延伸，并且实现自然通风。较小的中央单元配备了对流通风吊顶，部分穿越走廊。

42%的公寓位于东西向的建筑中，但是设计师将每个单元旋转了45度，并且按锯齿状排布。由于这个布局，所有东西向建筑中的单元实际上都是45度朝南。为了实现这一原则，建筑进深在有些地方达到25 m，这样可以在中部设置占地面积为5 m x 6 m的采光井和通风井。这些采光井和通风井将为所有的单元提供对流通风。

Reconstruction of Luan County Guangming Plaza
滦县光明商城改造

Keywords 关键词

Landmark Building
地标性强

Roof Garden
屋顶花园

Reasonable Layout
布局合理

Features 项目亮点

All the buildings are approx 100m height with completed space construction, have met the requirements of planning and construction which will be the landmark buildings of Luan County. Furthermore, by creating roof garden on the plazas, people in the apartments and residence can share with it.

建筑高度均接近 100 m，建筑空间构成完整，满足规划和开发建设成滦县地标要求。商业广场上设屋顶花园，公寓和住宅可以共同享有优美环境。

Location: Tangshan, Hebei, China
Architectural Design: Beijing Orient China Too Architectural Design Engineering Limited Liability Company
Land Area (Excluding the Roads): 16,659 m²
Planning Floor Area: 134,600 m²
Overground Floor Area: 105,000 m²
Under Ground Floor Area: 29,600 m²
Plot Ratio: 5.16
Green Coverage Rate: 15%

项目地点：中国河北省唐山市
设计单位：北京东方华太建筑设计工程有限责任公司
规划用地面积（不含道路）：16 659 m²
规划总建筑面积：134 600 m²
地上建筑面积：105 000 m²
地下建筑面积：29 600 m²
容 积 率：5.16
绿 化 率：15%

■ **Overview**

This project is located in the center of Luan County, Tangshan, with an advantageous geographical location, proper construction and convenient transportation. It is an urban complex project including commercial, residence, apartment, supermarket, cinema and so on.

■ **项目概况**

本项目位于唐山市滦县中心，地理位置优越，用地规整，交通方便。该项目为城市综合体，包含商业、住宅、公寓、大型超市、电影院等。

■ **Planning**

The architectures will set back 32m from the planned red line in the east, and create two main entrance plazas in the northeast and southeast respectively to ensure the flow distribution. In the west, the architectures will set up a leisure plaza, next to the pedestrian street, for rest and amusement. And will also set back 8m and 5m from the planned red line in the north and south respectively. To build the Guangming shopping mall, as a main part of the integrated multi-functional commercial facilities, around Yanshan Avenue and pedestrian street, it will create a prosperous commercial landscape in Luan County. There will have high-end residence in the north and south, and high-end apartments close to the main road, all with an independent entrance.

■ **总体布局**

规划建筑东侧退规划红线32 m，同时在东北和东南两个主要出入口处后退设置入口广场，以便于人流集散；西侧临步行街设置休闲广场，便于休息娱乐；北侧退红线8 m，南侧退红线5 m。以滦县商贸城为主力店的综合商业设施沿燕山大街和步行街布置，使之沿主要道路形成繁华的商业景观。南北侧分别布置高档住宅楼，靠近主干道一侧布置高档公寓，均有独立的出入口进出。

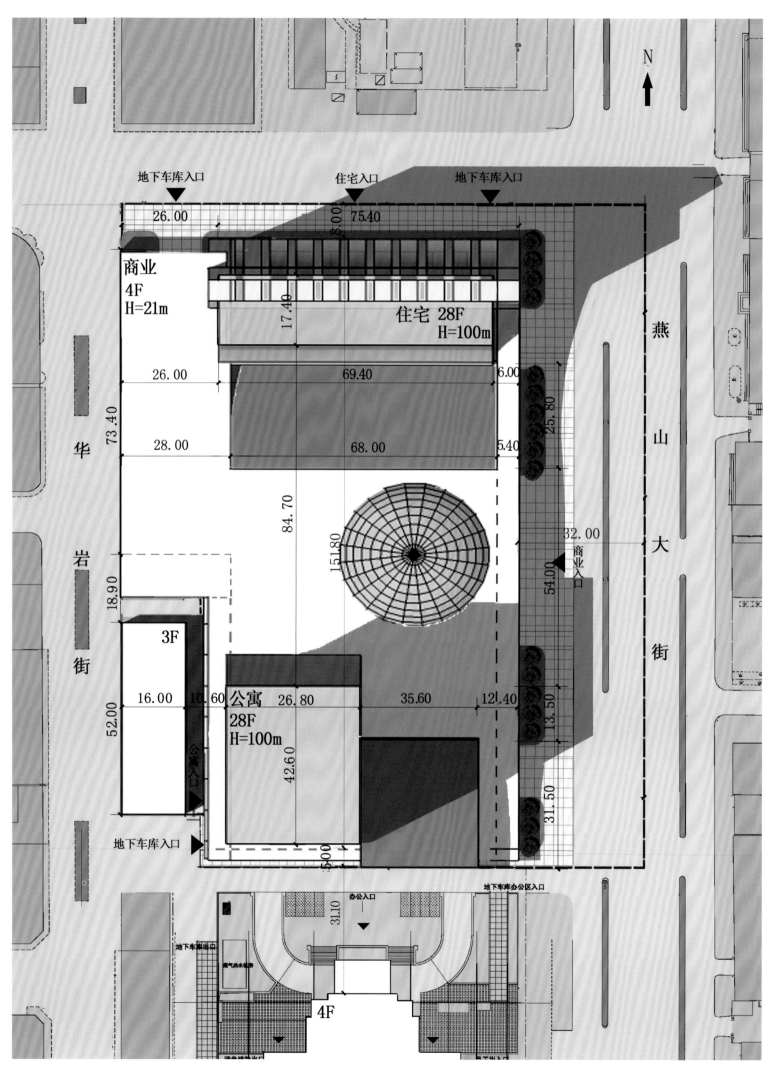

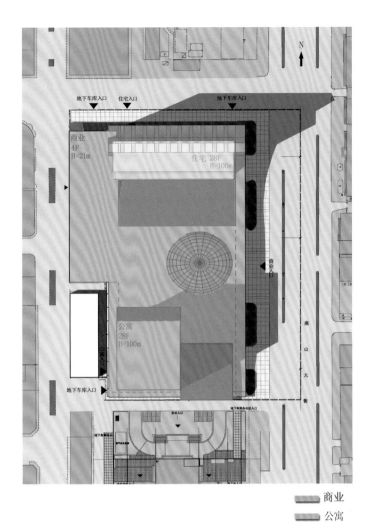

Functional Analysis Drawing 功能区分析图　商业　公寓　住宅

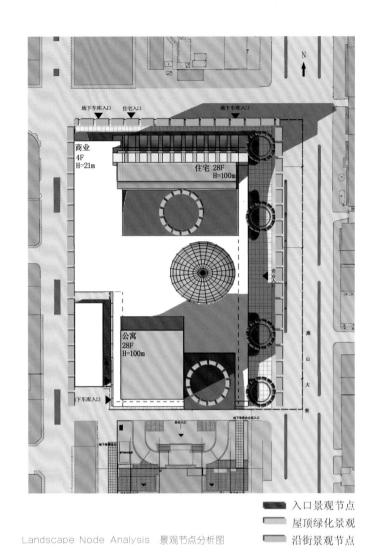

Landscape Node Analysis 景观节点分析图　入口景观节点　屋顶绿化景观　沿街景观节点

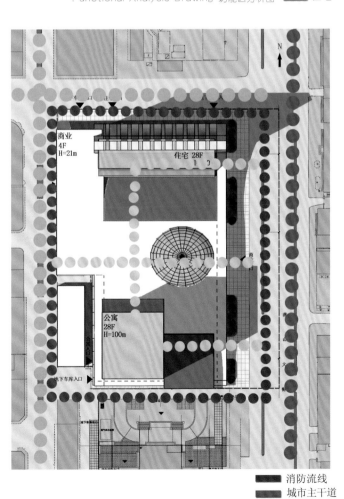

Traffic Network Analysis Drawing 交通流线分析图　消防流线　城市主干道　内街流线　步行街流线

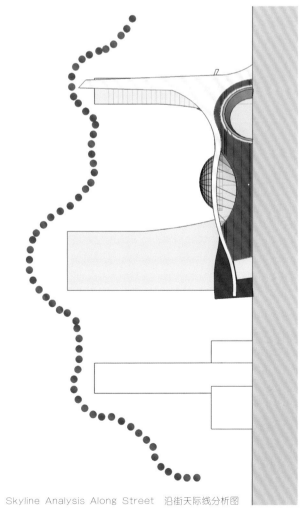

Skyline Analysis Along Street 沿街天际线分析图

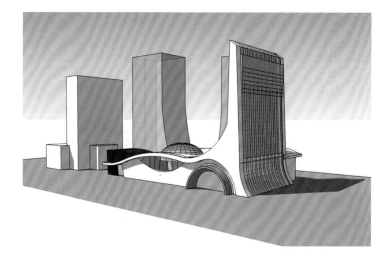

■ Monomer Building Design

Commercial Facilities

For this plaza, 2nd underground floor is garage, 1st underground floor is for the supermarket and store rooms, all the 1st to 4th floor on the ground are for commercials, also part of 4th floor have planned to be cinema and meeting rooms. Guangming shopping mall is in the northwest of the plaza, the rest will be the centralized business connecting by the commercial inner streets. No matter for renting for selling, it will be much better to have the commercial inner streets to connect all kinds of the stores and Guangming shopping mall, not only enrich the commercial inner space, but also make it more convenient for shopping. By creating an atrium with glass skylights at the bend of inner street in the 1st underground floor, the plaza have well lighting and environment.

Tower

There are two buildings for the tower, one is a 28-storey residential building, the other one is a 28-storey apartment block, both of them are approx 100m height with completed space construction, have met the requirements of planning and construction which will be the landmark buildings of Luan County. Furthermore, by creating roof garden on the plazas, people in the apartments and residence can share with it. In view of the scale and location of the residential building and perusing for the high level, the main house type for the residence is 120~1,150 m^2.

Basement One Plan of Commercial 商业地下一层平面图

Basement Two Plan of Commercial 商业地下二层平面图

■ 建筑单体设计

商业设施

地下二层为地下停车库，地下一层为超市和部分商业、库房，地上一、二层及三层、四层局部为商业，四层局部为电影院和会议厅。西北面为滦县光明商贸城，其余部分为以商业内街相连接的集中商业。无论出租或者出售，这将是更好的商业内街串联各个不同的店铺的购物中心，丰富了商业内部空间，也使购物更加方便，同时与滦县光明商贸城相连。内街地下一层转折处设中庭，顶部为玻璃天窗。广场采光好、环境优美。

高层建筑

高层建筑分二栋楼布置，其中一座为28层高的住宅楼，另一座为28层高的公寓楼。建筑高度均接近100 m，建筑空间构成完整，地标性强，满足规划要求和开发建设要求。商业上设局部屋顶花园，公寓和住宅可以共同享有优美环境。住宅以120～1 150 m^2/户的户型为主，考虑到住宅的规模和所处的地域，以追求高层次为主。

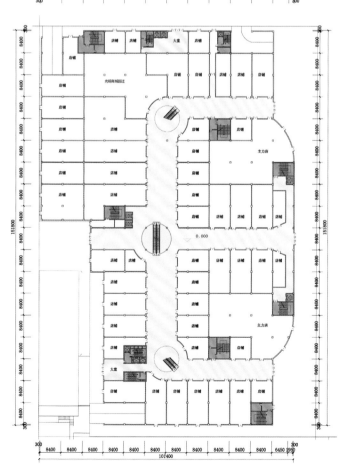

First Floor Plan of Commercial 商业一层平面图

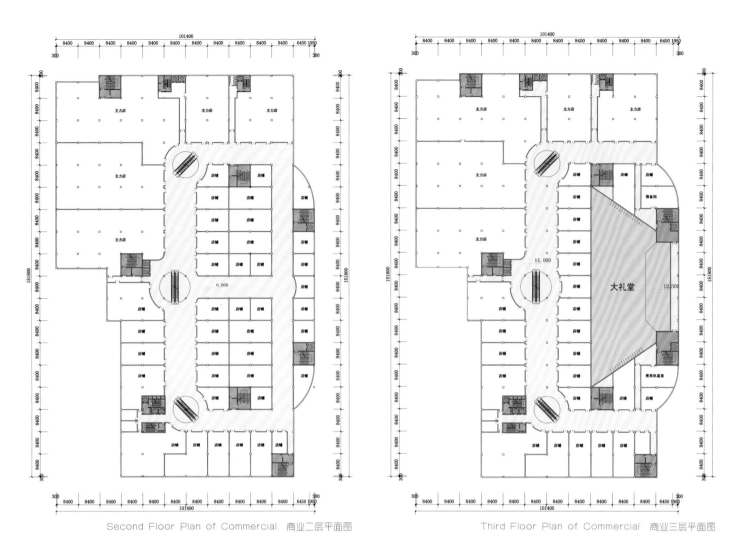

Second Floor Plan of Commercial 商业二层平面图

Third Floor Plan of Commercial 商业三层平面图

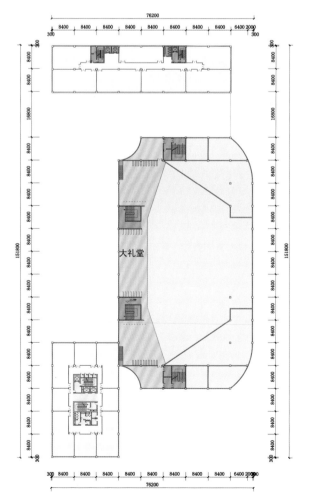

Fourth Floor Plan of Commercial 商业四层平面图

Fifth Floor Plan of Commercial 商业五层平面图

Quanshun Fortune Center Plot 14
泉舜财富中心 14# 地块

Keywords 关键词

Geometric Sculpture
几何雕塑

Concise Form
形体简洁

Hollow Glass
中空玻璃

Features 项目亮点

The building image bears strong sense of geometry sculpture, which looks tall and strong, and its shape and structure appear concise and modern, smooth and rich; its intensive vertical sense has strong feeling of strength, which is coordinate with the Chia Tai Super-tall Building.

建筑形象几何雕塑感强烈，挺拔有力，形体外观简洁而现代、流畅而丰富；建筑的竖向感加强，具有力度感，与正大超高层建筑协调。

Location: Luoyang, Henan, China
Architectural Design: Xiamen Hordor Design Group
Land Area: about 278,666.7 m²
Total Floor Area: 1,180,000 m²

项目地点：中国河南省洛阳市
设计单位：厦门合道工程设计集团有限公司
占地面积：约278 666.7 m²
总建筑面积：1 180 000 m²

■ **Overview**

Quanshun Fortune Center is a large urban complex including business, hotel, office and residence as one whole. This project design covers the hotel, property-type hotel, commercial building, office building and serviced apartments in on the northern side of plot 14. The project will become a landmark building improving the whole grade of Quanshun Fortune Center, and an important landmark along with plot 13 commercial complex in the future.

■ **项目概况**

泉舜财富中心是集商业、酒店、办公、居住为一体的大型城市综合体。此次设计项目为位于14#地块北侧的酒店、产权式酒店、商业、办公、酒店式公寓。本项目将是提升泉舜财富中心整体品位的标志性建筑，连同项目13#商业综合体未来将是该区域的一个重要地标性建筑。

■ **Design Concept**

The geographical analysis can indicate the importance and landmark attribute of the project. The project adopts concise, profound and distinctive modeling as the main design idea; it uses modern and contracted techniques to interpret the artistic conception of the construction, patterns of vertical lines to reflect the temperament of the tall building, and horizontal lines strew at random full of sense of rhythm; it seeks changes in contracted style to give the building modern cultural flavor.

■ **设计概念**

从区位分析，可以确定本项目的重要性和地标性。项目以简洁大方、造型独特为主要的设计构思理念；以现代简约的手法来诠释悠远的建筑意境，疏密有致的竖向线条体现了超高层建筑的气质，错落的横向线条镶嵌其中，充满了韵律感，在简约风格中寻求变化，使建筑呈现出现代的文化气息。

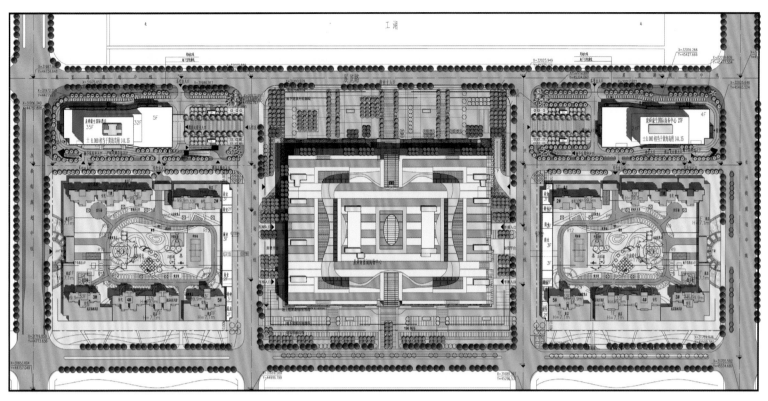

Site Plan 总平面图

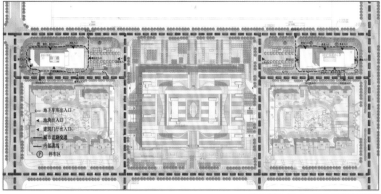
Traffic Drawing 交通分析图

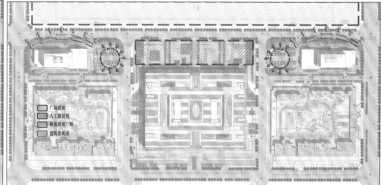
Landscape plan 景观分析图

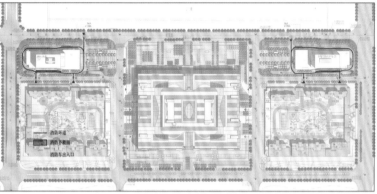
Fire Analysis Diagram 消防分析图

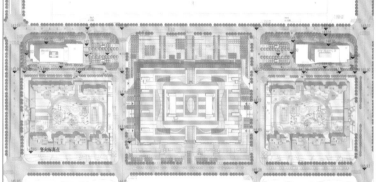
Vertical Analysis 竖向分析图

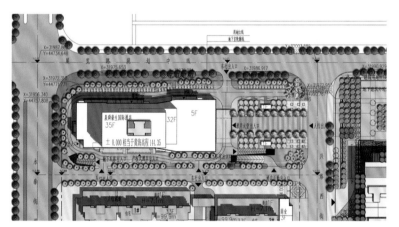
(14-1 North Plot) Overall Plan Rendering　(14-1 北地块) 总平面效果图

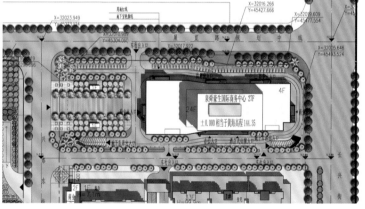
(14-2 North Plot) Overall Plan Rendering　(14-2 北地块) 总平面效果图

■ General Planning

Rich spatial levels: landscaping along the street—earth square—podiums—the main building—rear square—outer space levels of the fire channel.

The main building is set lateral around the base, and large enough urban space is set aside in the central part; there is urban landscape zone in the north side, directly facing Kaiyuan Lake, and its superior architectural image not only coordinates with the surrounding city skyline, but also reflects the temperament of the landmark building.

Maximized Viewing: The vast majority of the room and the main functional space can directly enjoy pleasant lake views.

Each partition has clear boundary and does not interfere each other. The entrances make a distinction between the important and the lesser one with strong identification and convenient access. The traffic flow line is clear and smooth that the plane and vertical transportation traffic are relatively independent but still can contact with each other.

The shape and outline of the building are well-proportioned and harmonious with the surrounding urban planning.

■ 总平面规划

空间层次丰富：沿街景观绿化—地上广场—裙房—主楼—后广场—消防通道的外部空间层次。

主楼设置在基地左右偏外侧，中部留出足够大的城市空间；北面留出城市景观带，直接面对开元湖，建筑形象优越，在与周围的城市天际线协调的同时体现地标性建筑的气质。

最大限度的观景：绝大多数房间和主要功能空间均可直接观赏怡人的湖景。

各部分分区明确，互不干扰。各入口主次分明，识别性强，出入便捷。交通流线清晰流畅，平面交通与垂直交通既自成体系，相对独立又能彼此联系。

建筑形体轮廓高低错落，虚实有致，与周边城市规划融为一体。

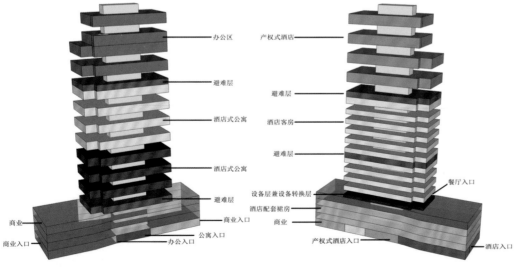

Vertical Function Analysis Diagram　立体功能分析图

■ Facade Modeling and Image Design

Skyline contour: On the premise of guaranteeing for the continuity and wholeness of the urban architectural skyline, it makes the building tall and straight with orderly ups and downs to form the main urban node in the Exhibition Road, stitching together to be an entire beautiful city construction movement.

Modeling image: The building image bears strong sense of geometry sculpture, which looks tall and strong, and its shape and structure appear concise and modern, smooth and rich; its intensive vertical sense has strong feeling of strength, which is coordinate with the Chia Tai Super-tall Building. Its concise podiums bond closely with the terrain, and coordinate with the commercial center facade of Plot 13.

Facade style: It takes modern and contracted style to shape rich architectural details, and adopts bright color tonal, which possesses age flavor. The project takes modern Chinese facade texture image to reflect the culture of the ancient capital, and creates an elegant and generous service image of the building as well.

Material design: It applies transparent colorless & blue grey hollow glass, and delicate metal materials to express the modern aesthetic feeling of the iconic building.

■ 立面造型及形象设计

天际轮廓：在保证片区城市建筑轮廓线的延续性与整体感的前提下，使该建筑挺拔，起伏有序，形成展览路上最主要的城市节点，共同联成整首美妙的城市建筑乐章。

造型意象：建筑形象几何雕塑感强烈，挺拔有力，形体简洁而现代、流畅而丰富；建筑的竖向感加强，具有力度感，与正大超高层建筑协调。简洁有力的裙房与地形结合紧密，与13#地块商业中心立面协调。

立面风格：采用现代简约风格，塑造丰富的建筑细部，采用较明亮的色调，具有时代气息。项目以现代中式立面肌理印象，体现古都的文化。同时营造出大厦高雅大方的服务形象。

材质设计：运用通透的无色中空玻璃与蓝灰色中空玻璃、精致的金属材料虚实结合表达标志性建筑的现代主义美感。

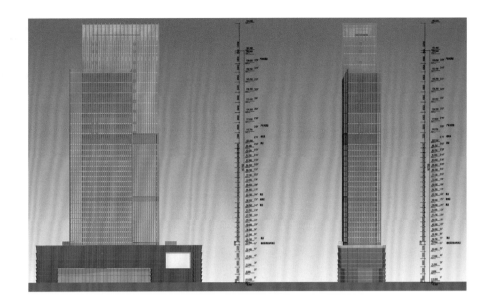

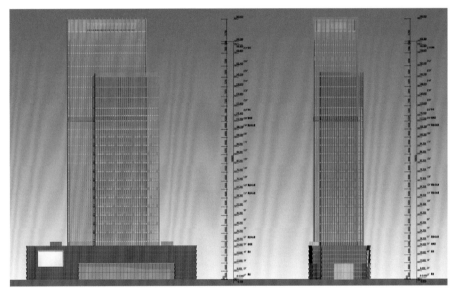

Lanyue Bay West Convention Centre & Mixed Development

揽月湾西地会议中心及综合发展

Keywords 关键词

- Curved Design 弧形设计
- High-rise Building 高层建筑
- Green Design 绿色设计

Features 项目亮点

The buildings are all curved to reflect the fluidity of the lake. The section of each tower is carved to house a large floating garden to give the master plan a softer, more resort like feel.

建筑物呈弧形设计，以反映湖泊的流动性。每座塔楼的剖面均以雕刻塑造大型海上漂浮花园，使总体规划显得更柔和，给人们度假村般的感觉。

Location: Changzhou, China
Client: GALAXY GROUP
Design Company: 10 DESIGN
Architect: Ted Givens
Architectural Design Team: Ted Givens, Adam Wang, Adrian Yau, Audrey Ma, Peby Pratama, Yao Ma
Land Area: 132,298 m²
Total Floor Area: 304,285 m²

项目地点：中国江苏省常州市
业　　主：星河集团
设计单位：拾稼设计
设 计 师：Ted Givens
设计团队：Ted Givens, Adam Wang, Adrian Yau, Audrey Ma, Peby Pratama, Yao Ma
占地面积：132 298 m²
总建筑面积：304 285 m²

■ Overview

The project is located in the West Tai Lake's Ecological and Leisure area, positioned to be one of the most important vacation resorts in the Delta Region of Yangtze River. The project will form a modern leisure and commercial community, providing an iconic luxury hotel, serviced apartments, conference centers, offices and public spaces.

■ 项目概况

项目位于常州市武进区滨湖新城的西太湖生态休闲区，将成为长江三角洲一个重要的旅游度假区。项目地块规划将形成一个现代化的休闲及商业社区，发展业态包括一间标志性的豪华酒店、酒店式公寓、会议中心、办公楼及公共空间等。

■ Design Inspiration

The master plan design was inspired by the local craft of bamboo carving and the aerodynamic shapes of modern boat design. The buildings are all curved to reflect the fluidity of the lake. A section of each tower is carved to house a large floating garden to give the master plan a softer, more resort like feel.

■ 设计灵感

此总体规划设计的灵感来自当地竹雕刻工艺及具有流线外形的现代船舶设计。建筑物呈弧形设计，以反映湖泊的流动性。每座塔楼的剖面均以雕刻塑造大型海上漂浮花园，使总体规划显得更柔和，给人们度假村般的感觉。

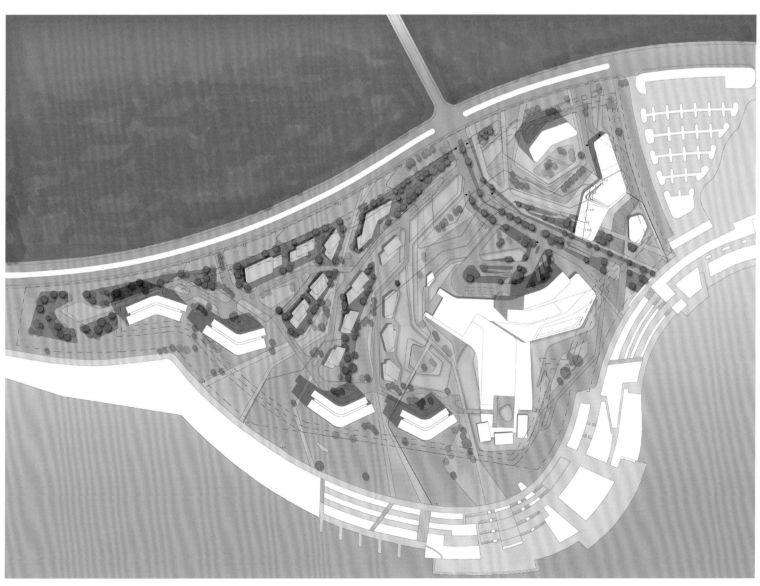

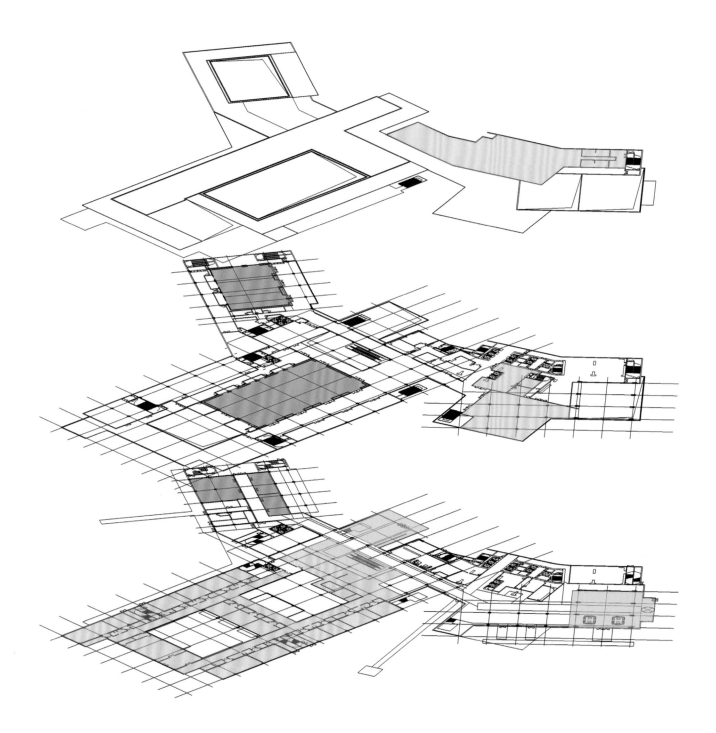

■ Design Goal

The MLP design is centered on 8 main design objectives:

1) To create an iconic high rise building facing the lake.

2) Arranging the buildings so they do not block the view of the lake from properties located behind/north of the site.

3) Minimize impact of shadows cast by towers on sites to the north of the MLP site.

4) To seamlessly integrate the retail and F+B facilities with the existing park to the east to help support the existing park and pull visitors into the MLP site.

5) Create a mixture of public spaces on the site including a large plaza, a large open park, an amphitheatre, and a series of playgrounds to support a variety of uses.

6) To provide each of the new residential units with views of the lake.

7) To maximize the possibility of cross ventilation for the buildings and to maintain lake breezes in all the garden areas.

8) To use the conference center as a sculptural object in the park, becoming a focal point for the entry sequence.

■ 设计目标

总平面的设计围绕8个主要的设计目标。

1.创建一个临湖的标志性高层建筑。

2.合理安排楼宇位置,避免其相互遮挡,以便使处于基地北部的楼宇均能享受到良好的湖景。

3.将建筑物的阴影遮挡减少到最小。

4.将基地内的餐饮购物设施与东侧的公园实现"无缝对接",既可以补充完善公共空间的服务设施,亦可将游客吸引至基地内。

5.创建具有综合功能的公共空间,包括大型广场、大型开放式公园、露天剧场以及游乐场所。

6.为每一个新建的住宅单位提供优美的湖景。

7.最大限度地保持建筑物的自然通风,使湖面的微风吹进花园的各个角落。

8.将会议中心作为园林中的雕塑元素,使其成为入口空间序列的视觉焦点。

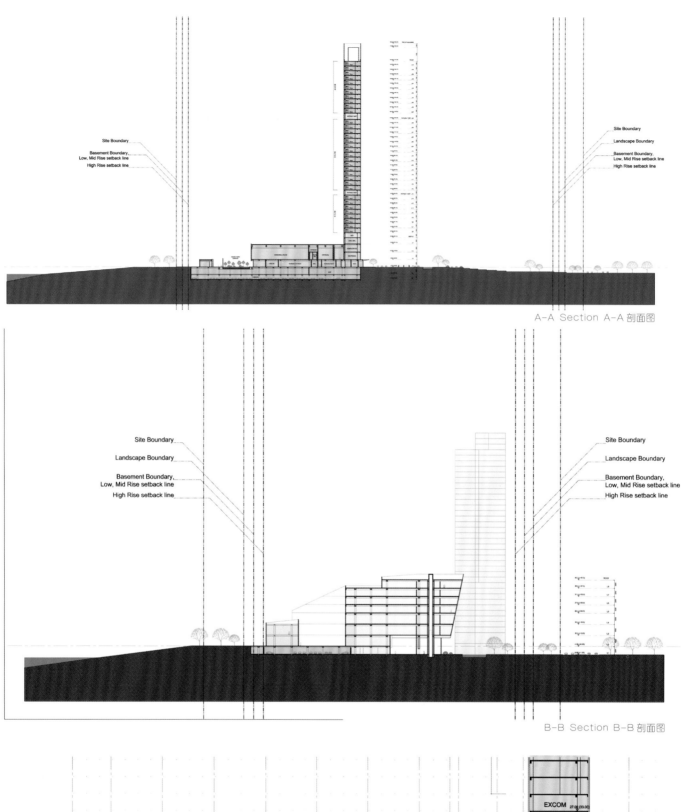

A-A Section A-A 剖面图

B-B Section B-B 剖面图

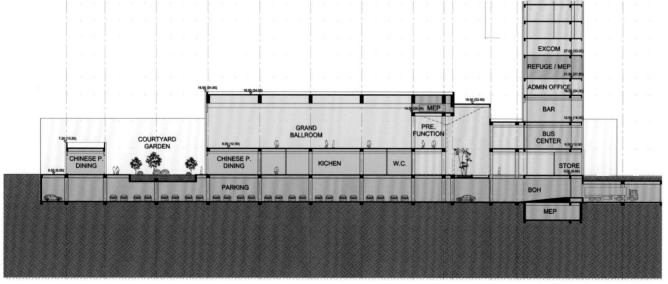

C-C Section C-C 剖面图

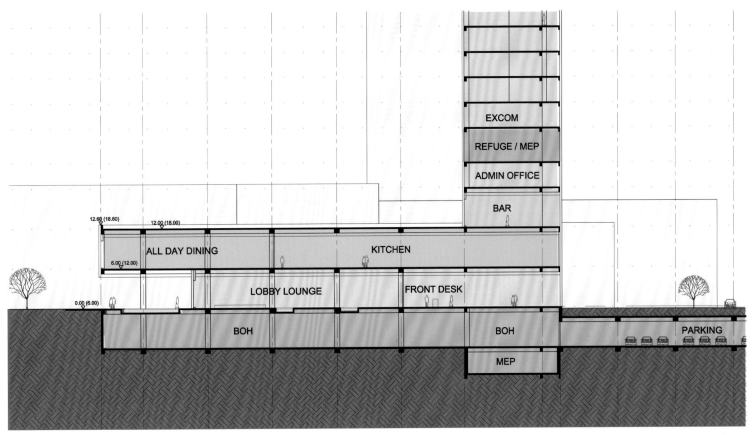

D-D Section D-D 剖面图

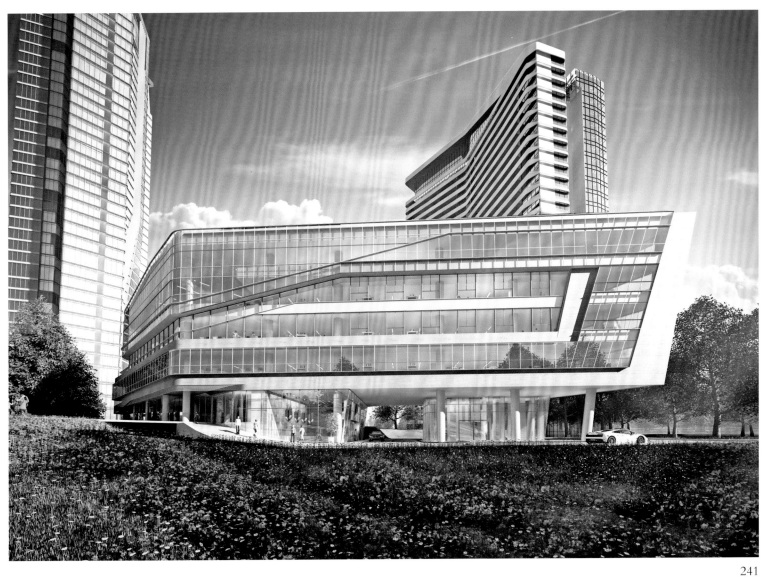

■ Planning & layout

Architectural design and planning arrangement respond both to its surrounding environment and the site itself. Careful consideration was paid to zoning and building arrangement within the site to maximize views and ventilation to all buildings, while at the same time retaining the existing view & breeze corridors. The concept of green design is further implemented via the introduction of green spaces above ground plane. Leisure facilities are designed on podium rooftops raised from the undulating landscape, while spaces are carved into the high-rise towers to form sky gardens for residents.

■ 规划布局

建筑设计及规划布局与周边环境及基地本身互相呼应，相得益彰。设计细心考虑了基地分区和基地内建筑物的布局，以获得最佳的景观效果及自然通风，同时亦保留现有景观及通风迴廊，通过进一步引入绿地空间彰显绿色设计的概念。休闲设施位于裙楼顶部，景观呈波浪形，高层塔楼间雕刻成的空中花园可供居民使用。

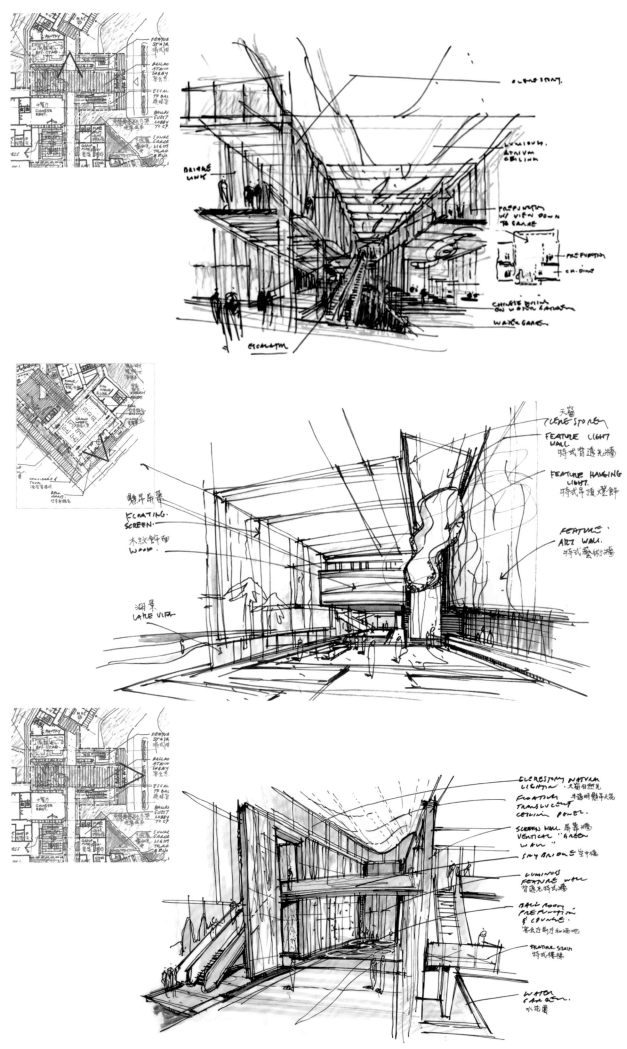

■ Architecture & Eco design

An iconic luxury hotel is located towards the east with maximum panoramic views to the lake. The form of the hotel sweeps down into the garden to create a conference facility, housed within a passive public park in the centre of the site. The conference centre will be a sculptural building acting as a focal point in the entry sequence into the site.

As an additional sustainable concept the hotel tower can be clad in a titanium dioxide nano-coating. The coating neutralizes air pollution and cleans the air. The tower can become an iconic symbol of the clean air of Changzhou.

■ 建筑与绿色设计

一栋标志性的豪华酒店位于基地东侧，坐拥最佳的湖景。酒店坐落在基地中心的公共花园，建筑外形从屋顶伸延至花园以形成一个会议中心。会议中心将作为一个雕塑建筑，使其成为入口视觉空间序列的焦点。

另一个可持续的绿色设计概念是将二氧化钛纳米涂料用于酒店塔楼外墙。该种涂料可以中和空气中的污染物，清洁空气。如果此概念得以运用，该酒店将会成为常州清洁空气的象征建筑。

■ Landscape design

Landscape design of the site forms the foundation of the overall planning and design of the masterplan. A mixture of public spaces including parks, plazas, an amphitheatre and playgrounds are designed to support a variety of uses. The nature of public usage of landscape areas allows the site to integrate seamlessly with the existing lake-side promenade to the east of the site.

■ 景观设计

项目的景观设计形成总纲发展蓝图的总体规划及设计的基础，公共空间集公园、广场、露天剧场及游乐场的设计具有多种用途。景观设计使基地与基地东侧的湖滨长廊配合得天衣无缝。

Guangzhou International Bioisland
广州国际生物岛

Keywords 关键词

"回" Layout
回字形布局

Eco Center
中央生态核

Green Design
绿色设计

Features 项目亮点

The architectural style is both conventional and free. The buildings stand in the periphery feature conventional outlines and multi layers with the blocks interweaving, the facades changing and the volumes rising. And the internal public spaces and transition spaces are open and well organized.

建筑造型设计是规整和自由的结合，外围主要功能用房外轮廓相对方正，通过体块的穿插，立面进退和建筑体量的高低变化形成丰富的层次，内部建筑的共享和联系空间外形自由流畅。

Location: Haizhu District, Guangzhou, Guangdong, China
Client: Guangzhou Development Zone Industrial Development Group Corporation
Architectural Design: Beijing CCI Architectural Design Co.,Ltd.
Site Area: 26,014 m²
Floor Area: 61,065 m²

项目地点：中国广东省广州市
业　　主：广州开发区工业发展集团公司
设计单位：北京中外建筑设计有限公司深圳分公司
用地面积：26 014 m²
建筑面积：61 065 m²

■ **Overview**

Located on Guangzhou Island of Haizhu District, Guangzhou International Bioisland (GIB) is developed to meet the need of biological technology development. This project is the first part of the development functioning as the production base, office and the research center.

■ **项目概况**

广州国际生物岛位于广州市海珠区官洲岛，一个以生物医药技术研究与产品开发为主的具有聚集效应的产业集群，是广州市发展高新技术产业的示范基地。本项目为生物岛的第一块进行开发建设的产业单元项目，使用功能由生产基地、办公、研发三部分组成。

■ **Planning**

The overall planning is inspired by the ideas of "combination of order and freedom", "central ecological core" and "sustainable and green building", forming a layout of two squares (one inside the other). In the center, it is the central eco park composed of water, wetland and square. The building is designed with conventional appearance and dynamic interior, providing sufficient spaces for mutual communication and information transfer, which will effectively enhance the cohesion of the enterprises.

■ **规划布局**

项目总体规划采用"规整和自由相结合"、"中央生态核"、"可持续发展的绿色建筑"等设计理念，保持建筑总体布局成回字形，中间是共享生态核——由水面、湿地和广场组成的中央生态园林。建筑造型设计采用外围规整、内部活跃的建筑布局，为人与人之间的交流和信息的传递创造了充足的空间，加强企业的凝聚力。

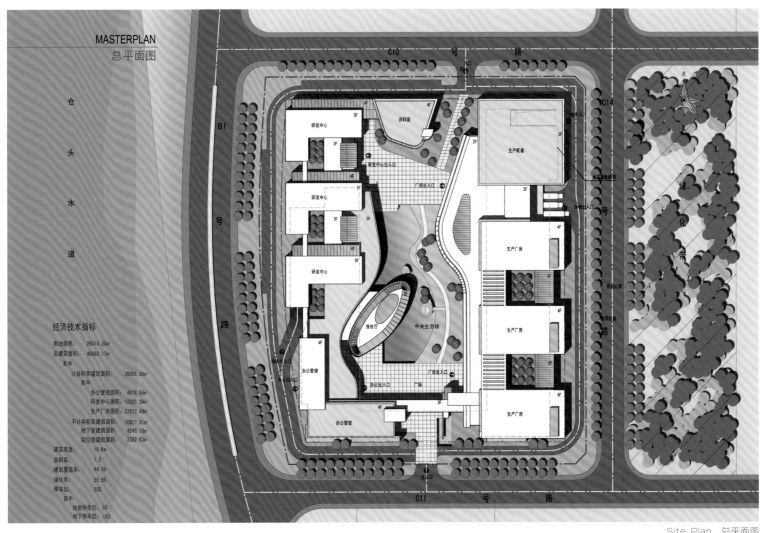

Site Plan 总平面图

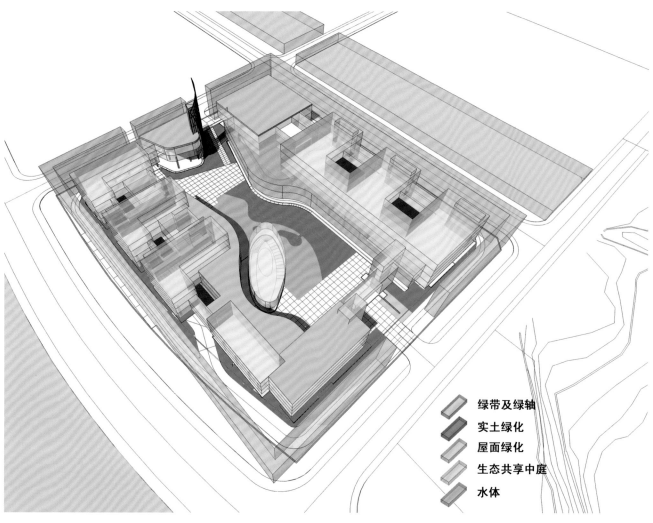

绿带及绿轴
实土绿化
屋面绿化
生态共享中庭
水体

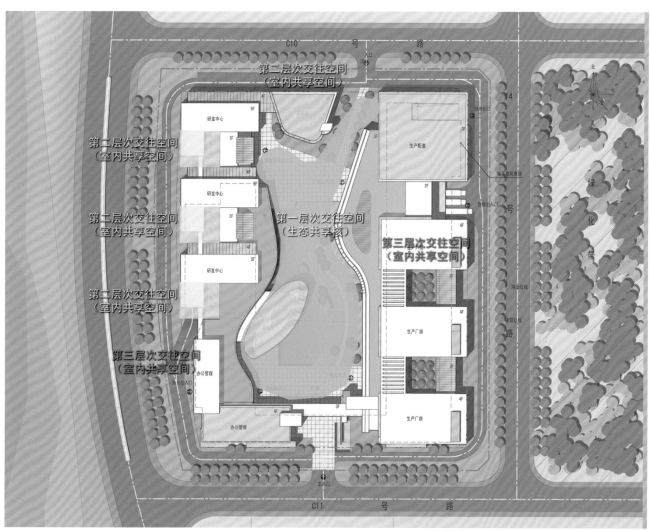

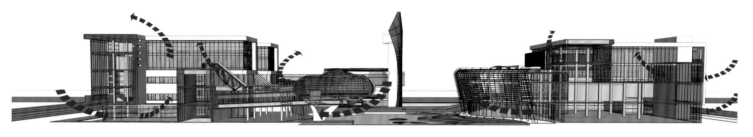

Architectural Design Description　建筑设计说明

■ Architectural Design

The architectural style is both conventional and free. The buildings stand in the periphery feature conventional outlines and multi layers with the blocks interweaving, the facades changing and the volumes rising. The internal public spaces and transition spaces are open and well organized. Paramecium-shaped lecture hall floating on the artificial lake, becomes the visual focus and symbolizes the most primitive state of life. It also echoes the subject of the biological research — the continuation and development of life.

■ 建筑造型设计

建筑造型设计是规整和自由的结合，外围主要功能用房外轮廓相对方正，通过体块的穿插，立面进退和建筑体量的高低变化形成丰富的层次，内部建筑的共享和联系空间外形自由流畅。草履虫形状的报告厅悬浮在中央人工湖上，成了视觉的焦点，象征最原始的生命状态——单细胞生物，突出生物科研产业的研究主题——生命的延续和发展。

■ Landscape Design

Open spaces and empty floors are left to bring the external landscape into the buildings. In the center, artificial lake, green belts and square are combined to form a central garden. Landscape gallery and the entrance hall are set to face to the central garden; two lower floors of the research building are set as ecological atriums which face to the central garden on the west side. Together with the internal courtyards between buildings and the roofs of different heights, it has formed a three-dimensional landscape system.

■ 景观设计

景观设计通过场地留空及建筑多层架空的手法形成对外的景窗引景入园；在场地的中央以人工湖、绿化带、广场等元素共同组成中心园林空间；生产基地面向中心园林一侧是南北伸展的景观长廊和入口大厅,科研用房的西侧面向中心庭院是两层通高的生态中庭，两者通过架空层、落地玻璃的方式把庭院景观引入室内；这三者加上建筑之间的内庭院和层层跌落的屋顶形成全方位立体化而且内外交融的景观格局。

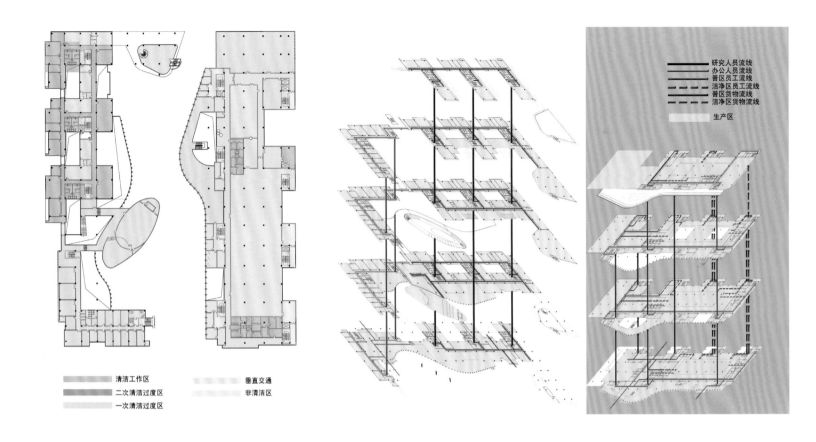
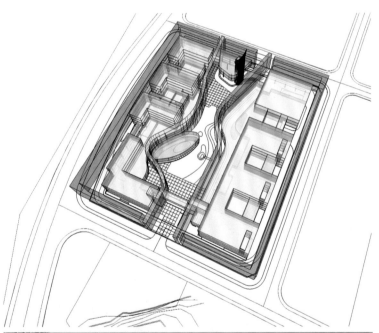

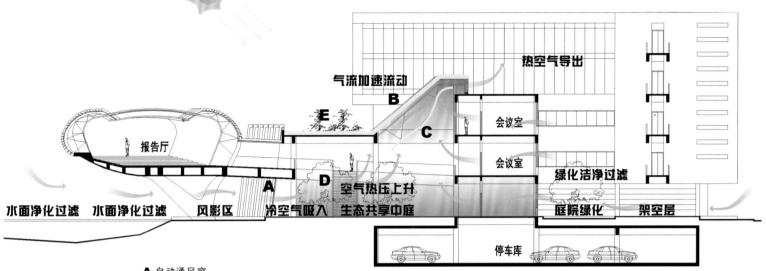

A 自动通风窗
B 特制热绝缘玻璃天窗
C 阳光放射板
D 中庭绿化
E 屋面绿化

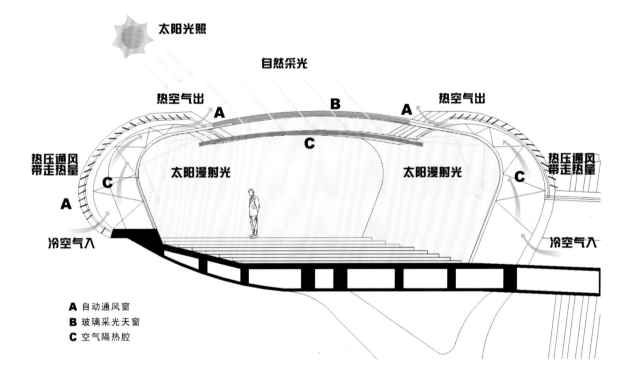

A 自动通风窗
B 玻璃采光天窗
C 空气隔热腔

Liuzhou Customs Port
柳州风情港

Keywords 关键词

LOW-E Glass
LOW-E 玻璃

Triangle Fold Surface
三角形折面

Added Facade
二次立面

Features 项目亮点

The structure design of the building is constituted by several changeable triangle fold surfaces, which is purposed to make three main public buildings spangle in the sun through the principle of light refraction, so as to achieve the effect of diamond tangent plane.

建筑本身的形体设计是由多个可变三角形折面所构成，目的是要使三栋主体公建通过光线折射的原理在日光中得以金光闪烁，从而达到钻石切面的效果。

Location: Liuzhou, Guangxi Zhuang Autonomous Region, China
Developer: Liuzhou Urban Construction & Investment Developing Co., Ltd.
Architectural Design: Guangzhou Hande Architectural Design
Total Land Area: 41,700 m²
Total Floor Area: 296,000 m²

项目地点：中国广西壮族自治区柳州市
开 发 商：柳州市城市投资建设发展有限公司
设计单位：广州汉德建筑设计事务所有限公司
总占地面积：41 700 m²
总建筑面积：296 000 m²

■ **Overview**

The project is located in the core zone of Liuzhou urban area, and it is an urban complex including residence, business, hotel, commercial buildings, Ming Dynasty City Wall relics, wax museum as well as urban square reconstruction, which is No.1 of the ten key projects in Liuzhou.

■ **项目概况**

项目用地位于柳州市城区的核心地带，该项目是一个包含住宅、商业、酒店、商务楼、明城墙遗址、蜡像馆，以及城市广场改造等的城市综合体，是柳州市十大重点项目之首。

■ **Design Concept**

This project gives response to the local region by strengthening the special geographical position of the project that it makes the architecture one of the earth landscape elements to harmoniously coexist and fully coordinate with the surrounding environment, and uses the shape of "stone" to remodel the earth texture and surface configuration in Liujiang River. Due to the special geographical position, the project design forges the hotel, office building and Ming Dynasty City Wall relics into three dazzling diamonds, which make the project look like as a pendant of a necklace, as if a falling pendant of Liujiang River after nightfall.

■ **设计立意**

本方案对于地域的回应，就是强化项目所在地的特殊地理位置，把建筑融为大地景观的元素，与周边环境和谐共生，充分交融，利用"石头"形态重塑柳江的大地肌理与地表形态。由于项目所嵌合的地理位置特殊，在项目设计时将酒店、写字楼和明城墙遗址打造成三颗璀璨夺目的钻石，使得项目如同项链上的链坠，入夜后成为柳江之坠。

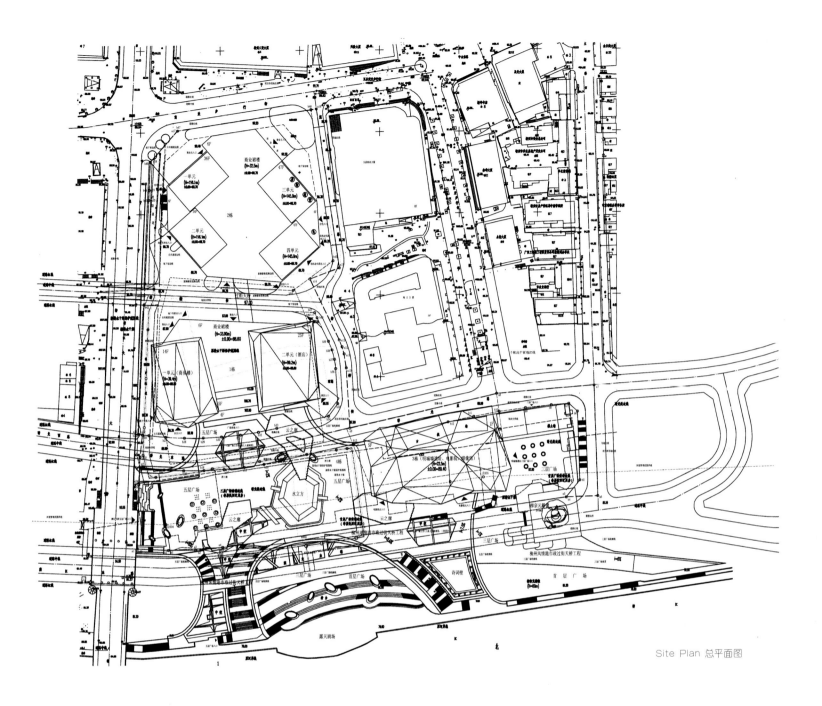

Site Plan 总平面图

■ Architectural Design

The structure design of the building is constituted by several changeable triangle fold surfaces, which is purposed to make three main public buildings spangle in the sun through the principle of light refraction, so as to achieve the effect of diamond tangent plane. The building perimeter protection materials chose golden hollow LOW-E glass, which not only ensures good day-lighting for the building, but also can avoid pollution phenomenon caused by large area of glass curtain wall light reflextion, and has effective energy saving.

■ 建筑设计

建筑本身的形体设计是由多个可变三角形折面所构成，目的是要使三栋主体公建通过光线折射的原理在日光中得以金光闪烁，从而达到钻石切面的效果。建筑外围保护材料选用了金色中空LOW-E玻璃，既保证了建筑物良好的采光，又避免了大面积玻璃幕墙光反射所造成的光污染现象，同时也能有效地节能。

■ Publicly Constructed Housing

From the perspective of "function decide the style", housing and public construction are different from each other; housing appears a good deal of balconies, windowsills, split-air-condition external hanging shelves, etc., which will form many "points" on the facade side and is not conducive to a complete surface form. So the design takes the method of adding facade to hide different "points" behind the "surface", so as to make the facade graceful and entire publicized effect. Aluminum materials are arranged in parallel "line" to form a whole large facade, and the through-hole aluminum plates become a diagonal slash to divide the facade, which also echoes with the diamond slash in front.

■ 住宅公建化处理

住宅与公建的不同，如果从"功能决定样式"的角度出发来区分，住宅会出现大量的阳台、窗台、分体空调外挂位置等。而这些会在立面构图上形成许多的"点"，不利于完整面的构成。本设计通过二次立面的设置处理，将不同的"点"隐藏在"面"的背后，让立面形成大气整体的公建化效果。铝质材料以"线"式并向排列，形成整体的大立面，通孔铝板形成划分立面的斜线，与前方的钻石斜线形成呼应。

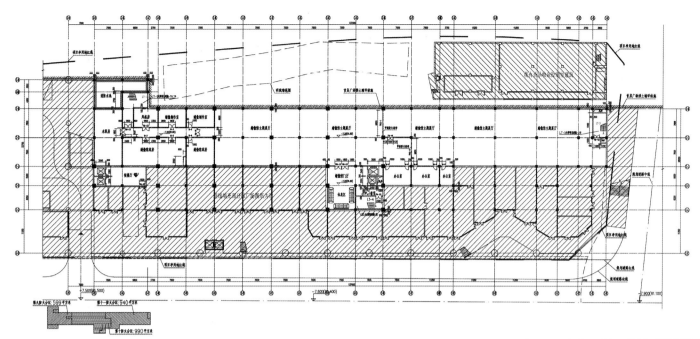
Basement One Plan 地下一层平面图

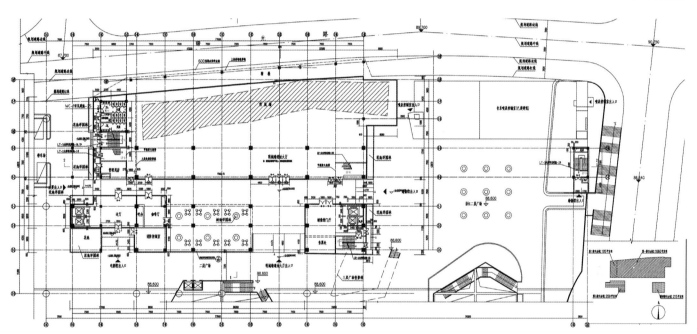
Ground Floor Plan 首层平面图

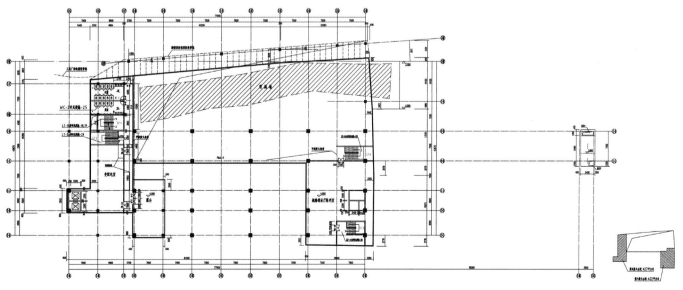
Groundfloor Mezzanine Plan 首层夹层平面图

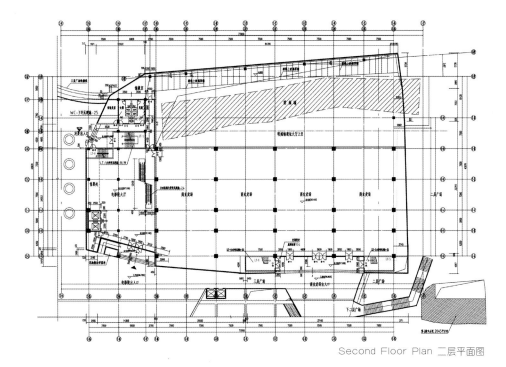

Second Floor Plan 二层平面图

Third Floor Plan 三层平面图

Fourth Floor Plan 四层平面图

■ Plaza Details

The plaza design makes full use of the three-dimensional terrain of the base with three layers. The first layer is used for Commercial Street and Riverside Park, the second floor is designed to be the citizens' leisure square and the top layer services as a major festival activity field.

The framework of three layers strides and reels across each other with fluent, concise and graceful lines, and the curved design goes coherent. There is falling water in the plaza across the three layers, which can reflect its dimensional feeling and changeable features, and skillfully combines the three-layer plaza as a whole. The water falls down from the top of a "water cube" glass viewing sketch on the third floor. On the other side, the poems on the "poem wall" by Liu Zongyuan enable the citizens in the plaza to enjoy the poetic charm and imposing manner. Besides, the designers use the detailing design of the poem wall to make it become the day-lighting surface for the next layer of the plaza.

■ 广场细部

广场的设计充分利用了三层基地的立体高差地形。广场的首层为商业街和滨江公园，二层设计成市民的休闲广场，而顶层则作为大型节庆活动的场所。

三层的架构相互跨绕，线条流畅、简洁而大气，弧形的设计一气呵成。在广场上另外设计了一条跨越三层的落水很能反映出它的立体感和多变的特色，并且巧妙地把三层的广场贯穿为一体。落水从位于三层的一个名为"水立方"的玻璃观景小品顶部飞流直下。另一侧的柳宗元"诗词壁"的诗词让广场周边的市民能领略到诗的神韵和气势。同时设计师还利用诗词壁的大样设计，使其又成为下一层广场的采光"天面"。

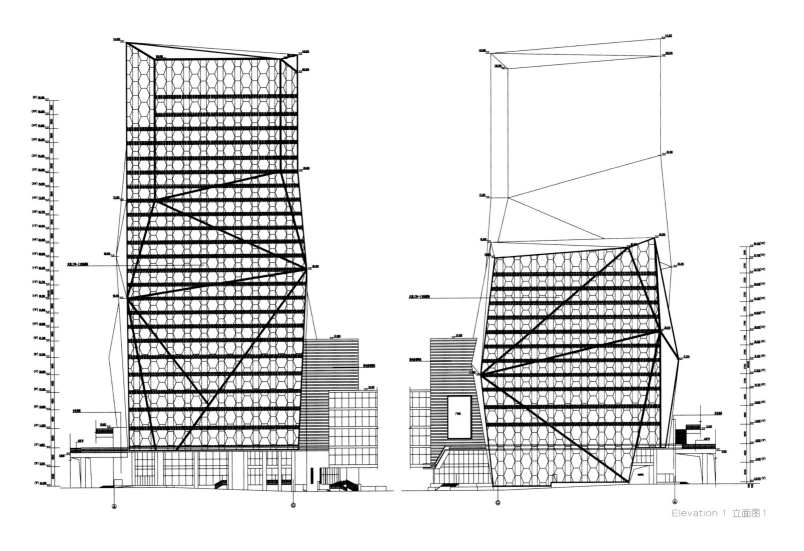

Elevation 1 立面图1

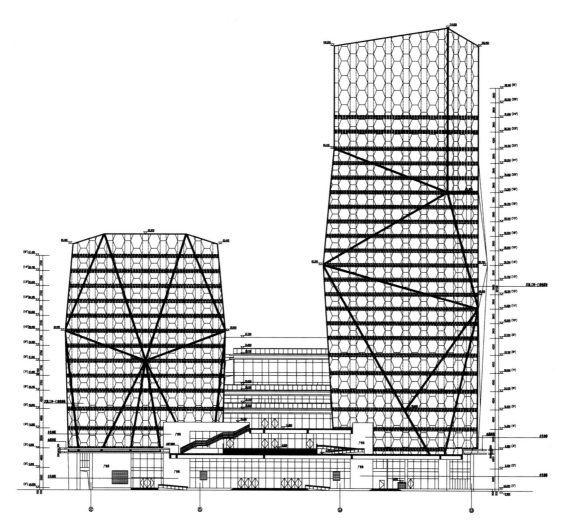

Elevation 2 立面图 2

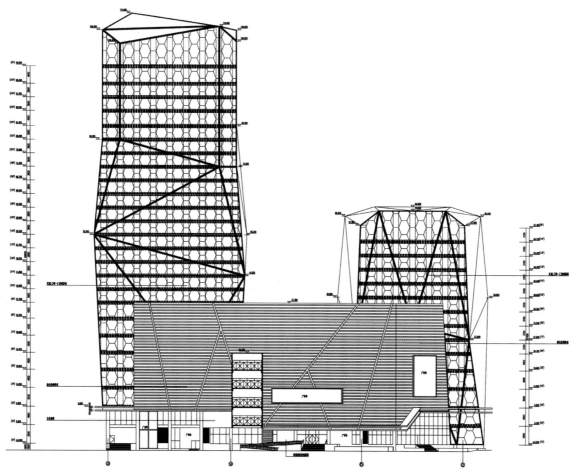

Elevation 3 立面图 3

CITIC Jinluan Bay, Zhangzhou
漳州中信东山岛金銮湾

Keywords 关键词

- **Level Design**
  层次分布
- **Dynamic Axis**
  活力轴线
- **Folded Hotel Building**
  折线形酒店

Features 项目亮点

Buildings along the commercial street are designed in semi-open courtyard-style. They are in small size with big distance between buildings to get maximum landscape views and avoid sight interference.

商业街单体采用半开敞小院落的形式，建筑小巧精美、间距较大，保证了基地与自然交流的顺畅和观海视线的通透。

Location: Zhangzhou, Fujian, China
Developer: CITIC (Fujian) Investment Co., Ltd.
Architectural Design: Beijing CCI Architectural Design Co., Ltd. (Shenzhen)
Site Area: 61,697 m^2
Floor Area: 98,272 m^2

■ Overview

The project is located on Dongshan Island of Zhangzhou, at the south end of Fujian Province. Dongshan Island looks like a flying butterfly, thus it is also called "Butterfly Island". According to the planning, there are four parts along the coastline: seaside commercial street, waterfront villas, resort hotel and seaview apartments. The main landscape avenue extends to the apartment buildings from the beaches, forming a dynamic axis that connects the buildings with the beautiful surroundings.

项目地点：中国福建省漳州市
开 发 商：中信福建投资有限公司
设计单位：北京中外建筑设计有限公司深圳分公司
用地面积：61 697 m^2
建筑面积：98 272 m^2

■ 项目概况

项目位于福建省漳州市的东山岛，地处福建省南端，东山岛状似翩翩起舞的蝴蝶，素有"蝶岛"的美称。本项目分为四个部分，沿海岸线往内依次是滨海商业街、亲水别墅、度假酒店及海景公寓。主景观道由海滩深入公寓腹地，成为活力轴线，将整个项目与曼妙的自然美景连为一体。

Site Plan 总平面图

■ Architectural Design

Buildings along the commercial street are designed in semi-open courtyard-style. They are in small size with big distance between buildings to ensure great landscape view and sea view. Villas are arranged in leaf vein pattern in four clusters to get maximum landscape view and avoid sight interference. Fold-line shaped hotel building is full of the atmosphere of holiday and intrest. Landscapes in semi-open courtyard combine together with the buildings to provide great views ofeach room. The lobby is set on the central axis of the site to create multi-level landscape. Apartment buildings and garden houses are arranged in order to provide rhythmic skyline which also makes the project a beautiful landscape in this area.

■ 建筑设计

商业街单体采用半开敞小院落的形式，建筑小巧精美、间距较大，保证了基地与自然交流的顺畅和观海视线的通透。别墅分为四个组团，叶脉状排布，争取最大的景观视野，同时避免了相互之间的视线干扰。折线形酒店则充满了度假感和趣味性，半围合的庭院景观与建筑融为一体，而且每间客房都有良好的观海面。大堂更是位于基地中轴线上，景观层次丰富悠远。公寓和花园洋房的错落有致使得天际线表现出灵动的韵律感，也令项目本身成为了区域内一道优美的风景。

Fantasia · Long Nian International Center
花样年·龙年国际中心

Keywords 关键词

- Situational Shopping Circle 情景式购物圈层
- One-stop 一站式
- Stylish 别具风情

Features 项目亮点

Huasheng Town, which belongs to Long Nian International Center, is a unique situational shopping circle and stylish one-stop consumption space that invested and operated by Fantasia, and includes well-known brands like Sheraton Hotel and Jackie Chan Yaolai International Cinema.

花样年·龙年国际中心所属商业——花生唐，是由花样年统一招商运营，由喜来登酒店及成龙耀莱影院等国际一线品牌商家领军，打造独特的情景式购物圈层，拥有别具风情的餐饮娱乐休闲配套，创造"一站式有趣、有味购物消费空间"。

Location: Chengdu, Sichuan, China
Developer: Chengdu Wangcong Real Estate Development Co., Ltd.
Land Area: 126,666.7 m²
Floor Area: 800,000 m²
Plot Ratio: 4.01
Green Coverage Ratio: 35%

■ **Overview**

Located at the center of "Fantasia · Wangcong Cultural Park" in southwest Pixian County, Chengdu, Long Nian International Center is adjacent to provincial cultural relics protection units "Wangcong Temple" and "Wangcong Cultural Park" central park. The project accommodates mainstream boutique department store, world-renowned star hotels, supermarkets, theme commercial district, landmark-class office building, LOFT, housing and so on, which is another masterpiece sponsored by Fantasia company.

项目地点：中国四川省成都市
开 发 商：成都望丛房地产开发有限公司
占地面积：126 666.7 m²
建筑面积：800 000 m²
容 积 率：4.01
绿 化 率：35%

■ **项目概况**

花样年·龙年国际中心位于成都郫县县城西南部的"花样年·望丛文化园"片区腹地，紧邻省级文物保护单位"望丛祠"及"望丛文化园"核心园区。项目包括主流精品百货、全球知名星级酒店、超市、主题商业街区、地标级写字楼、LOFT、住宅等多种业态，属花样年公司打造的又一城市综合体力作。

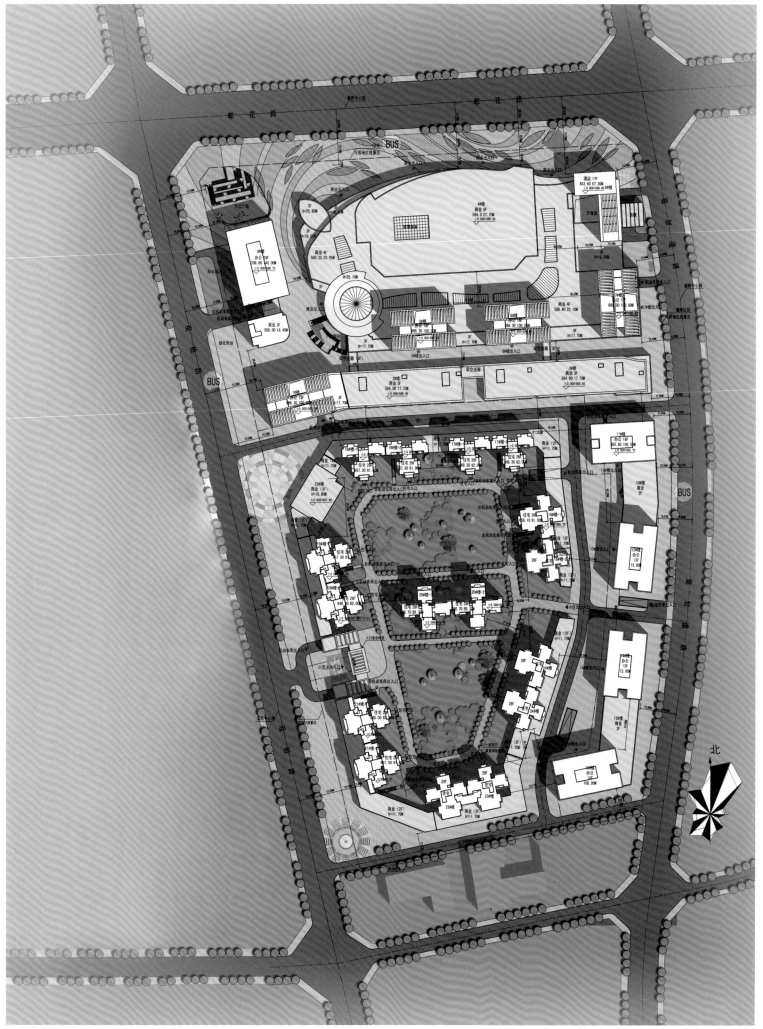

Site Plan 总平面图

■ Situational Shopping Circle

Huasheng Town, which belongs to Long Nian International Center, is a unique situational shopping circle and stylish one-stop consumption space that invested and operated by Fantasia, and includes well-known brands like Sheraton Hotel and Jackie Chan Yaolai International Cinema. Once completed, the project would promote the urban construction vigorously, and improve living and commercial facilities and become the new center of Pixian.

■ 情景式购物圈层

花样年·龙年国际中心所属商业——花生唐，是由花样年统一招商运营，由喜来登酒店及成龙耀莱影院等国际一线品牌商家领军，打造独特的情景式购物圈层，拥有别具风情的餐饮娱乐休闲配套，创造"一站式有趣、有味购物消费空间"。项目建成后，将大力推动片区城市建设，完善片区生活及商业配套，成为郫县城市新中心。

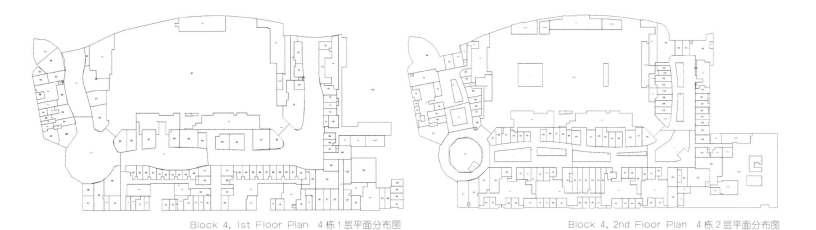

Block 4, 1st Floor Plan　4栋1层平面分布图

Block 4, 2nd Floor Plan　4栋2层平面分布图

Block 4, 3rd Floor Plan　4栋3层平面分布图

Block 4, 4th Floor Plan　4栋4层平面分布图

Shijiazhuang Urban Complex
石家庄中委城市综合体

Keywords 关键词

Function Division
功能分区

Unified Color
色彩统一

Void Atrium
中庭挑空

Features 项目亮点

Color used on the facade is overall considered, e.g., the soil color on the office building gradually turns into yellow on the hotel and aparment. A part of the large ramp, which leads to the building, forms a huge public plaza connecting office area and residential area.

外立面的颜色使用上是整体考虑，从办公楼泥土的颜色渐变到酒店、住宅的黄色。通向大楼的大型坡道的部分形成一个巨大的公共广场连接办公和住宅部分。

Location: Shijiazhuang, Hebei, China
Architectural Design: WEAVA Architects

项目地点：中国河北省石家庄市
建筑设计：法国韦瓦建筑设计公司

■ Overview

Located on the southwest side of the 2nd Ring Road in Shijiazhuang City, the project was designed to create a new commercial landmark in this area. In terms of building orientation, it aims to get the maximum daylighting while minimizing the side effects of the shadow of high-rise buildings on the surrouding residences.

■ 项目概况

该项目位于石家庄市二环路的西南边，致力于打造该地区的新型商业地标。从坐落方向上考虑，该方案旨在获得最大的采光面同时尽量减少高层建筑的阴影对周围房屋的影响。

■ Facade Design

Compared to apartment tower and office biulding, hotel tower is the highest among the three buildings. A unified complex is formed thanks to skyline and the same materials. Facade unity is achieved in different ways according to different functions. How open should the facade be depends on the privacy of the buildings, that's why the office tower is more open and the hotel is relatively modest and has clear symbolic room division.

Color used on the facade is overall considered, e.g., the soil color on the office building gradually turns into yellow on the hotel and aparment. A part of the large ramp, which leads to the building, forms a huge public plaza connecting office area and residential area. Void atrium is arranged in the commerical area under the hotel and office building, making it a more flexible space.

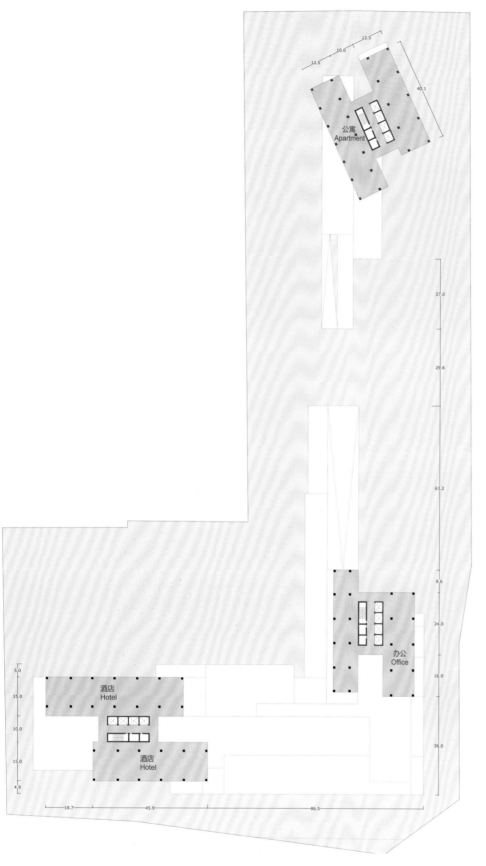

Floor Plan 平面图

■ 外立面设计

相比公寓塔和办公大楼，酒店塔为三座塔中最高的。通过天际线以及相同的材料，使综合体形成统一。外立面上则依据使用功能的不同，以不同的方法来达到统一，外立面开放的程度也取决于建筑的私密程度，所以这就是为什么办公塔比较开放，而酒店则相对收敛，且有清晰标志性的客房划分。

外立面的颜色使用上也是整体考虑，从办公楼泥土的颜色渐变到酒店、住宅的黄色。通向大楼的大型坡道的部分形成一个巨大的公共广场连接办公和住宅部分。酒店及办公下方的商业部分设有中庭挑空，使得空间上更加灵活多变。

■ Hotel Design

Hotel is located on the west end of the site. It is divided into two parts by the central axis of the regional center. Star height is kept within 100m in accordance with building codes. The gap on the top floor can meet the use of hanging garden. Guests can get into the hotel directly on the west side where there is a huge cantilevered glass is used as an obvious entrance sign.

■ 酒店设计

酒店坐落于地块的最西边。大楼被该区域中心形成的中轴线分成两部分。在城市建造规范的限制下，楼梯高度均保持在100 m以内。顶层的落差可满足空中花园的使用。客人可以从西侧直接进入酒店，巨大的悬吊伸出的大型玻璃体入口是明显的入口标志。

■ Commercial Center Design

Commercial center is a transition between the hotel and office building. It occupies the largest area, up to four storeys that acomodates retail stores, leisure services and so on. It is broken by a large three-storey high covered walkway which is used as an internal street for the mall. It also extends into the rest area at the bottom of the existing residential area and office building.

■ 商业中心设计

连接酒店和办公楼的过渡部分是商业中心。它占据的面积最大，多达4层楼层，满足零售商铺、休闲及其他综合使用功能。商场被一条巨大的三层高的有顶步行通道打断，该通道同时也兼为商场的内部街道。商业和休闲功能同样延伸扩展至其余现有住宅区底层的休息区和办公大楼中去。

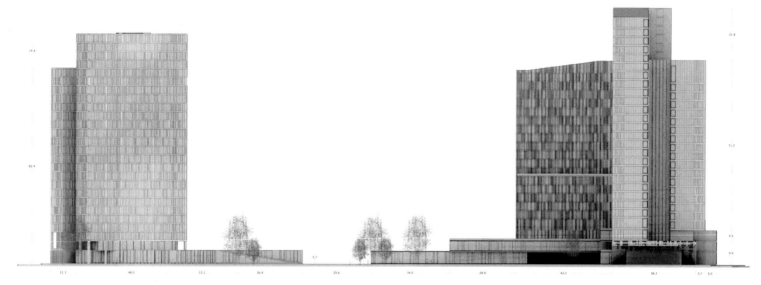

West Elevation 西立面图

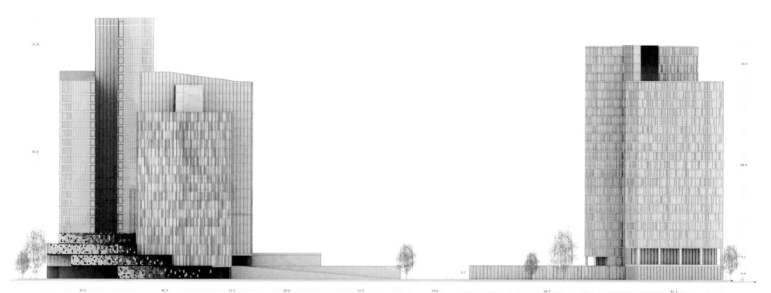

East Elevation 东立面图

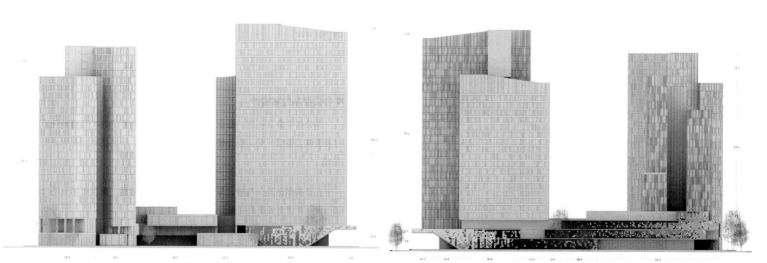

North Elevation 北立面图　　　　　　　　　　South Elevation 南立面图

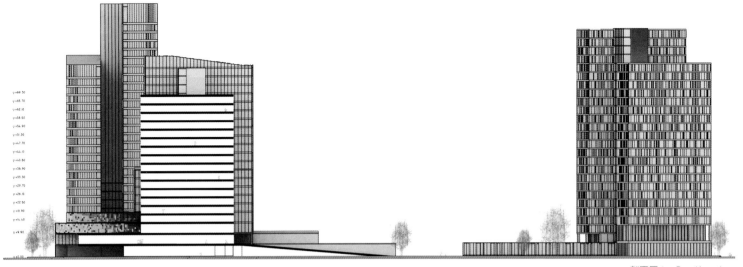

剖面图1 Section 1

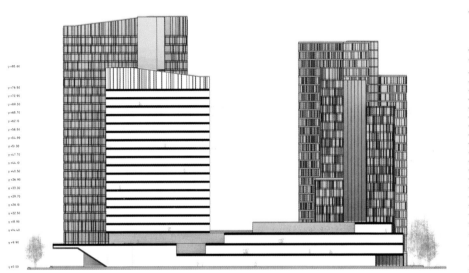

剖面图2 Section 2

剖面图3 Section 3

■ **Office Building & Apartment Design**

The office building is also divided by the central axis like the hotel. It is relatively lower than the other two towers. Office entrance is on the north side that allows the commercial center to extend in.

Apartment building is in the north part where there is more privacy and inclusive. The rest part connects a large public space through independent passgeway.

■ **办公楼与公寓设计**

办公楼与酒店一样，在中轴线上被划分开来，酒店的高度相对于另两座塔楼的高度来说要低一些。办公楼的入口坐落在北边，使主楼的商业中心可沿街延续至此。

北边的部分是公寓楼，那里更具私密性和包容性。剩余的部分通过独立的通道连接一个大型的公共空间。

Chengdu Lingdi International Plaza
成都领地国际广场

Keywords 关键词

Twin Tower
双子塔楼

Arc-shaped Connection
弧形连接

Grand Volume
体量宏大

Features 项目亮点

In order to break the traditional design method, designers connect the bottom of the twin tower in an arch shape with the concept of "the eye of the city", showing an extraordinary appealing sense of sculpture.

塔楼采用双子塔的布局方式，为打破传统的设计手法，设计师以"城市之眼"为立意，将塔楼下部以弧形连接，使建筑呈现出雕塑般的非凡吸引力。

Location: Chengdu, Sichuan, China
Developer: Chengdu Lingyue Real Estate Development Co., Ltd.
Architectural Design: Sichuan Cendes Architecture Engineering Design Co., Ltd.
Land Area: 9,246 m²
Floor Area: 187,279 m²
Plot Ratio: 5.9
Green Coverage Ratio: 15%

项目地点：中国四川省成都市
开 发 商：成都领悦房地产开发有限公司
建筑设计：四川山鼎建筑工程设计股份有限公司
占地面积：9 246 m²
建筑面积：187 279 m²
容 积 率：5.9
绿 化 率：15%

■ Overview

Located in West Chengdu (west entrance), this project was to create a multi-functional international city complex. It seeks to establish a vibrant and dynamic new landmark in accordance with "northward improvement" that the government is calling for, and makes due contribution to improve the urban form of northern Chengdu.

■ 项目概况

项目用地位于成都西大门的门户位置，定位为具有多重功能的国际化城市综合体。项目力求为成都打造出一个充满朝气和活力的新地标，响应政府号召加快成都"北改"，为改善成都北部片区的城市形态作出应有贡献。

■ Architectural Design

In order to break the traditional design method, designers connect the bottom of the twin tower in an arch shape with the concept of "the eye of the city", showing an extraordinary appealing sense of sculpture. Succinct folded form on the annex building not only meets the needs of commercial function but echoes with the tower perfectly.

■ 建筑设计

塔楼采用双子塔的布局方式，为打破传统的设计手法，设计师以"城市之眼"为立意，将塔楼下部以弧形连接，使建筑呈现出雕塑般的非凡吸引力。裙房简洁的折面形态，既满足商业功能的需要，又和塔楼形成完美的呼应。

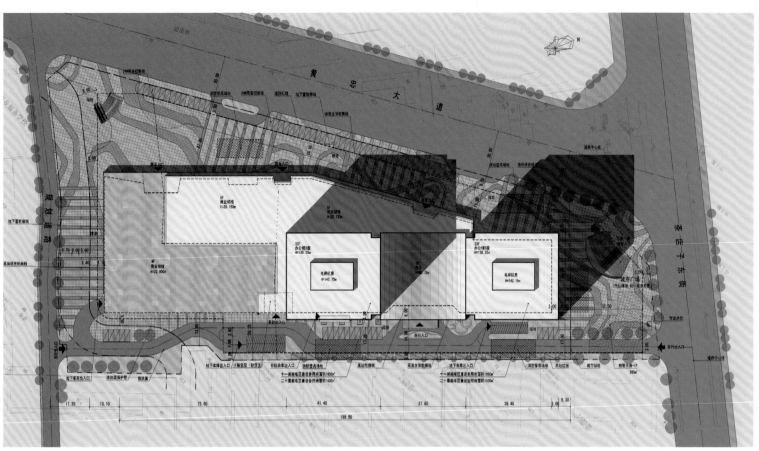

Site Plan 总平面图

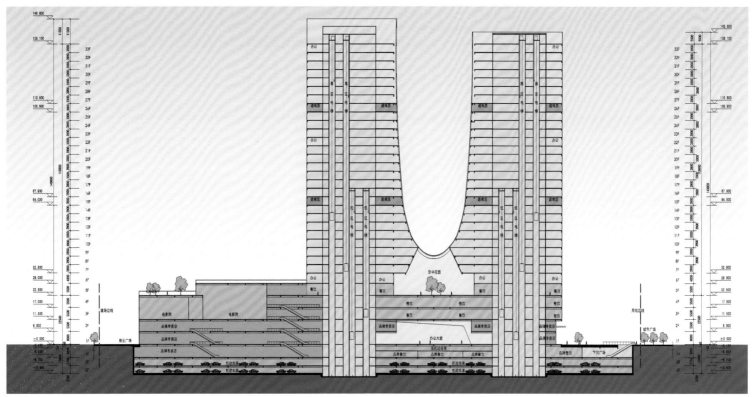

Section 1-1　1-1 剖面图

Front Elevation 正立面图　　　　　　　　　　End Elevation 侧立面图

281

Administrative Center in Fengdong New Town, Xixian New Area

西咸新区沣东新城管委会政务中心

Keywords 关键词

Streamline Shape
流线型

Unique Appearance
外观独特

Fifth Facade
第五立面

Features 项目亮点

The streamlined shape breaks the sense of solemnness and restriction the rectangular site expresses, extends the streamlined element to the fifth facade—the roof and even affects the use of the material, forming a rich office service space and brand image of unique taste.

建筑形体上采用流线型整体设计，打破矩形场地所带来的严肃和约束，并将流线这种元素抽象延伸到建筑第五立面——屋顶的设计，甚至阐释在材料的运用上，形成丰富的办公服务空间和独具品味的品牌形象。

Location: Xi'an, Shaanxi, China
Developer: Xi'an Fengwei Real Estate Co., Ltd.
　　　　　Poly Real Estate Group Co., Ltd.
Architectural Design: Xi'an Baian Architectural Design Co., Ltd.
Total Floor Area: 59,047 m²

项目地点：中国陕西省西安市
开 发 商：西安沣渭地产有限公司
　　　　　保利房地产（集团）股份有限公司
设计公司：西安百岸建筑设计有限公司
总建筑面积：59 047 m²

■ Project Planning

The project was designed to extend the urban axis and make a creative symbolic space for social interaction through strengthening systematic structure. A required sense of core is established by controlling the view of the surroundings. The city square is defined to lead people to enter a multi-functional office service environment that presents multi-sights.

■ 项目规划

规划上通过延续城市轴线关系，强化系统结构，将城市间不同片段之间建立起一个具有创新性的标志空间提供社会性交流互动，并控制建筑在周边环境范围内的视野来营造所需的核心感，并界定由城市广场引领人们进入和展示多重景观和多种体验的办公服务环境。

■ Design Concept

In considering the structure relationship between Fengdong New Town and Xixian New Area & Xi'an and the spatial interaction system between various functional areas, designers proposed spatial core sense as the theme and created a thriving urban image through analyzing and reorganizing the building and the site which present the sustainability of resources and environment, thus to highlight the position and image of the project in the process of urban planning.

■ 设计理念

考虑了关于沣东新城与西咸新区以及大西安的结构组成关系，以及项目内各功能区域的空间互动体系，提出以空间核心感为主题，通过建筑与场地的解析和再组织，创造出蓬勃发展的城市意象，展现新城资源、环境整合利用的可持续发展性，从而强调项目在城市规划发展中的核心位置与门户形象。

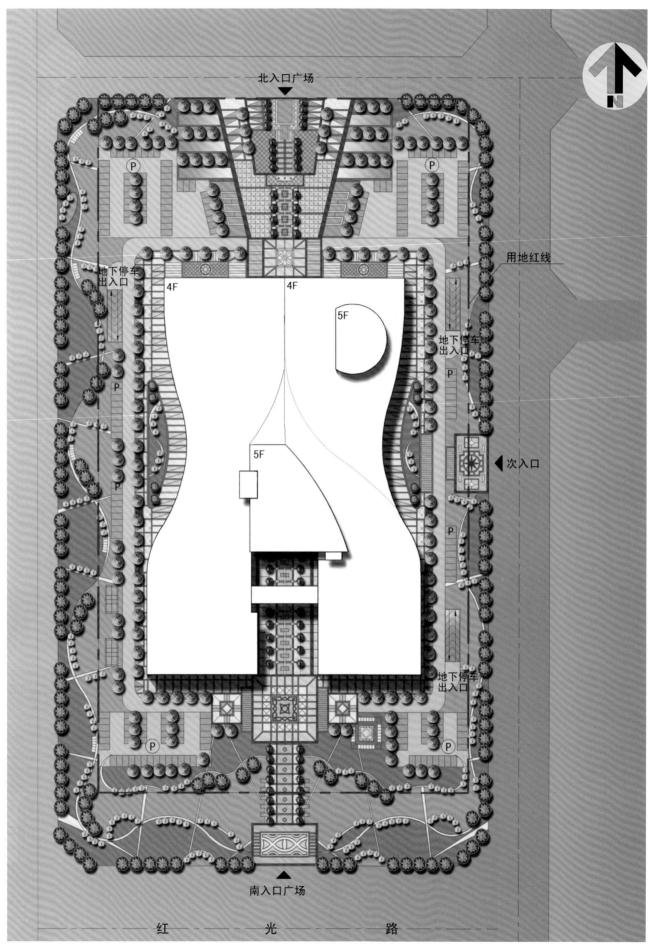

Site Plan 总平面图

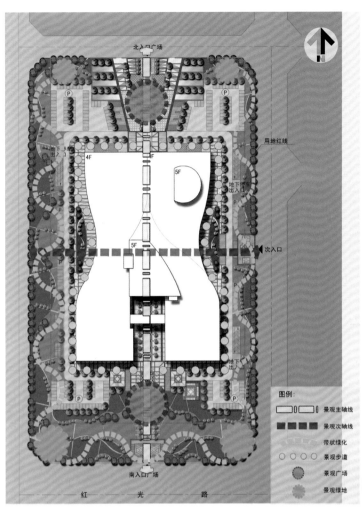

Landscape Analysis Drawing
景观分析图

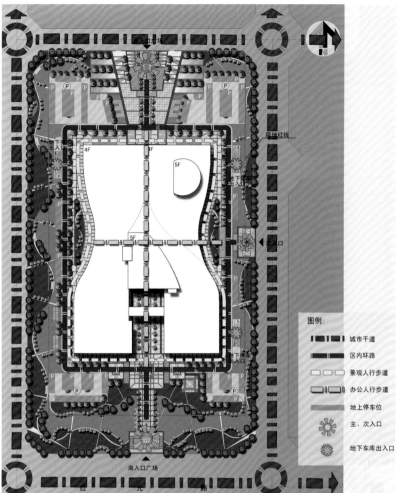

Traffic Analysis Drawing
交通分析图

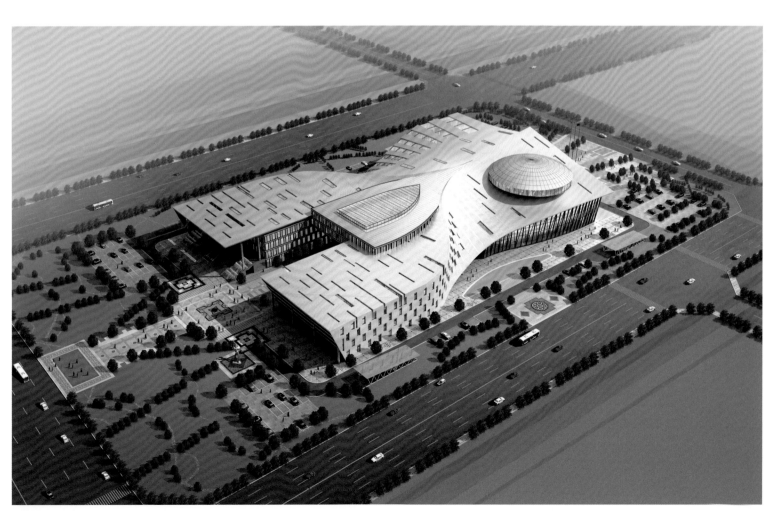

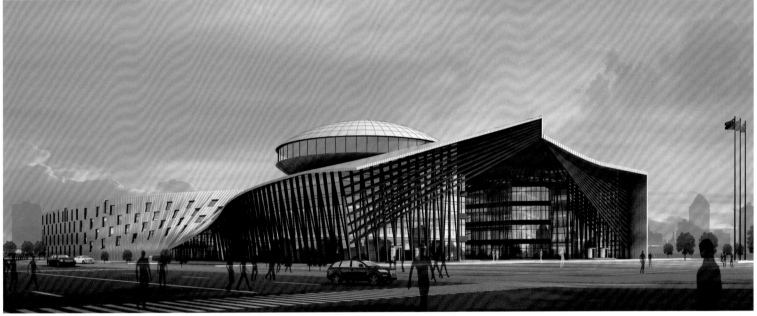

■ Functional Layout

The administrative center is a comprehensive office building accommodates office, exhibition and service. Generally the height is controlled within four floors, and there are five floors partially. Function space is arranged vertically and connected by horizontal passages, which ensures the relative independence between different flow lines without losing the interactive space, providing a flexible office environment.

■ 功能布局

政务中心是集办公、展示、服务等为一体的综合办公服务建筑，总体高度控制在四层，局部五层。功能空间沿竖向垂直分布，横向通道将其贯穿连接，既保证不同行为组织流线之间的相对独立性，又不失交流互动空间，提供了一种具有灵活协调性的办公环境。

■ Architectural Design

The streamlined shape breaks the sense of solemnness and restriction the rectangular site expresses, extends the streamlined element to the fifth facade—the roof and even affects the use of the material, forming a rich office service space and brand image of unique taste and creating innovative, high-end yet humanistic urban image.

■ 建筑设计

建筑形体上采用流线型整体设计，打破矩形场地所带来的严肃和约束，并将流线这种元素抽象延伸到建筑第五立面——屋顶的设计，甚至阐释在材料的运用上，形成丰富的办公服务空间和独具品味的品牌形象，创造出具有创新性、高端性且不失人文关怀的城市意象。

Civil Aviation Air Traffic Control Base Logistics Service Center
民航空管基地后勤服务中心

Keywords 关键词

Reasonable Planning
规划合理

Terrain-oriented
顺应地势

Independent Group
独立组团

Features 项目亮点

In terms of architectural form, designers strive to place its planning in an important position while pursing reasonable, convenient and personalized spatial structure. In addition, height different is flexibly, and buildings are properly arranged in accordance with terrain in the premise of minimum land transformation and earthwork.

建筑形态本案在力求空间规划结构合理、便捷和个性化的同时，将建筑形态规划放到一个重要位置。基地内现状地形具有一定的高差，设计中巧妙利用现有高差，在对地形改造最小、土方量最小的前提下合理布局建筑，使建筑顺应地形的走势。

Location: Xi'an, Shanxi Province, China
Developer: Hesheng Civil Aviation Air Traffic Control Construction &Development Co., Ltd.
Architectural Design: Xi'an Baian Architectural Design Co., Ltd.
Beijing Zhonghan International Architectural Design Co., Ltd.
Total Floor Area: 332,500 m²

项目地点：中国陕西省西安市
开 发 商：陕西合盛民航空管建设开发有限公司
设计公司：西安百岸建筑设计有限公司
　　　　　北京中翰国际建筑设计有限公司
总建筑面积：332 500 m²

■ Overview

Located in Wudong Village, Pangguang Town, Xi'an, the project land is about 38 km from Xi'an Downtown. Civil Aviation Air Traffic Control Base Logistics Service Center is dedicated to year-round recreation & short-term vacation for senior people in civil aviation system and other services like conference reception and staff training.

■ 项目概况

项目地块位于西安市户县庞光镇乌东村，距西安市中心约38 km，民航空管基地后勤服务中心主要为民航系统老年人提供常年休闲养老和短期度假疗养，兼具会议接待、员工培训等服务功能。

■ Design Concept

Characterized by holiday recreation, the project takes the advantages of convenient regional transportation (a short distance to Xi'an Downtown) to provide people a destination for short-term tour. In its unique cultural space, one may find a rest and restorative space to make him harmonious, tranquil and relaxed both mentally and physically.

■ 设计理念

项目设计以休闲度假为特色，针对该区域与西安便捷的交通、近距离的优势，为人们提供一个短期旅游的好场所。其独特的文化空间，营造出一个可以令精神和身体达到和谐、宁静、放松和忘我的休憩天地和疗养空间。

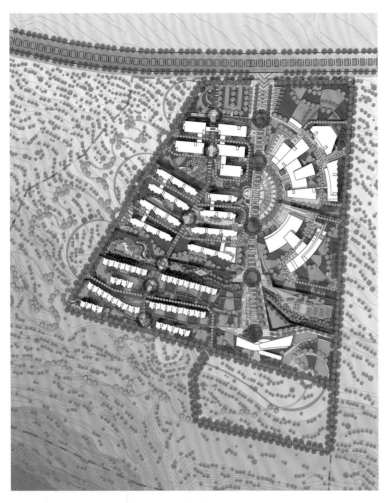

Traffic Analysis Drawing
交通分析图

Landscape Analysis Drawing
景观分析图

■ Layout

Overall planning starts from analyzing the land, and then incorporating natural elements, function blocks, life tracks, sight and landscape into a unified system, which makes the land layout, road planning and spatial organization more logical to combine with the base. Since the base is divided into several plots in different sizes, designers try to create some independent yet interconnected groups. Each group has its own road system and echoes with each other in architectural form and architectural style.

In terms of architectural form, designers strive to place its planning in an important position while pursing reasonable, convenient and personalized spatial structure. In addition, height different is flexibly, and buildings are properly arranged in accordance with terrain in the premise of minimum land transformation and earthwork.

■ 规划布局

总体规划从整体用地的分析入手,将自然要素、功能区块、生活轨迹、视线景观等纳入统一体系,使小区的用地布局、道路规划、空间组织等与基地的结合更具逻辑性。由于基地被地势分隔成大小不等的几块,因此规划尽量使各个地块形成相互独立又相互关联的几个组团,每个组团内道路自成系统,各组团通过建筑形态和建筑风格的统一和呼应来取得整体效果。建筑形态本案在力求空间规划结构合理、便捷和个性化的同时,将建筑形态规划放到一个重要位置。基地内现状地形具有一定的高差,设计中巧妙利用现有高差,在对地形改造最小、土方量最小的前提下合理布局建筑,使建筑顺应地形的走势。

■ Landscape Design

Combining with the existing base terrain features, landscape design incorporates leisure space with ornamental space, mountain views, waterscape and building space together to form a continuous theme landscape concept.

■ 景观设计

景观设计结合基地现有的地形特征将休闲空间与观赏空间相互穿插,和山景、水景、建筑空间融合在一起,形成连续的主题景观概念。

Langxi Guogou City
郎溪国购城

Keywords 关键词

Beautiful Building
建筑优美

Reasonable Layout
布局合理

Green Design
绿色设计

Features 项目亮点

As a major feature of the project's design, the "green community" is planned to be efficient, comfort, energy-saving, environment-friendly and ecological balanced. By creating a close-to-green space, residents could make the most of the green area.

营造"绿色社区"是本设计的一大特点，以求达到高效舒适、节能环保、生态平衡的目的。亲绿空间的创造，最大限度地使人进入绿地。

Location: Xuancheng, Anhui, China
Architectural Design: CCI Architecture Design & Consulting Co., Ltd.
Plot D1
Site Area: 84,040.5 m²
Total Floor Area: 126,537 m²
Plot D2
Land Area: 69,998 m²
Site Floor Area: 208,321 m²

项目地点：中国安徽省宣城市
设计单位：上海新外建工程设计与顾问有限公司
D1地块
用地面积：84 040.5 m²，
总建筑面积：126 537 m²
D2地块
用地面积：69 998 m²
总建筑面积：208 321 m²

■ **Overview**

Langxi Guogou City is located in the core area of Langxi County. The project is divided into two plots as D1 and D2, where D1 is designed for commercial use by a 4-storey large department store, a 3-storey mid-size commercial building, three 6-storey commercial complex buildings and 10 commercial units of 2-3 layers, while D2 is used as residential area for 21 high-rise resident buildings.

■ **项目概况**

基地位于郎溪县核心区，分为D1和D2两个地块：D1地块为商业用地，由1栋4层大型百货、1栋3层中型商业大楼、3栋6层商业综合楼、10栋2~3层商铺组成；D2地块为住宅用地，由21栋高层住宅组成。

■ **Project Orientation**

Based on Langxi core area's locational advantages, future commercial core strengths and landscape advantages, the project strives to build D1 and D2 plots into a one-stop shopping center including large commercial buildings, mid-size commerical buildings, commercial complex buildings plus shops along the street, and also a high-end humanities community with a beautiful landscape, elegant architectures, noble character, complete supporting facilities as well as rich residential atmosphere.

■ **项目定位**

项目将依托郎溪县核心区的区位优势、即将成型的商业核心优势和地块内的景观优势，在D1地块打造一个包含大型商业、中型商业、综合商业楼、沿街商铺等在内的综合一站式购物中心；在D2地块打造一个景观优美、建筑典雅、品格高尚、配套设施完善、居住氛围浓厚的中高档人文社区。

总平面图 Site Plan

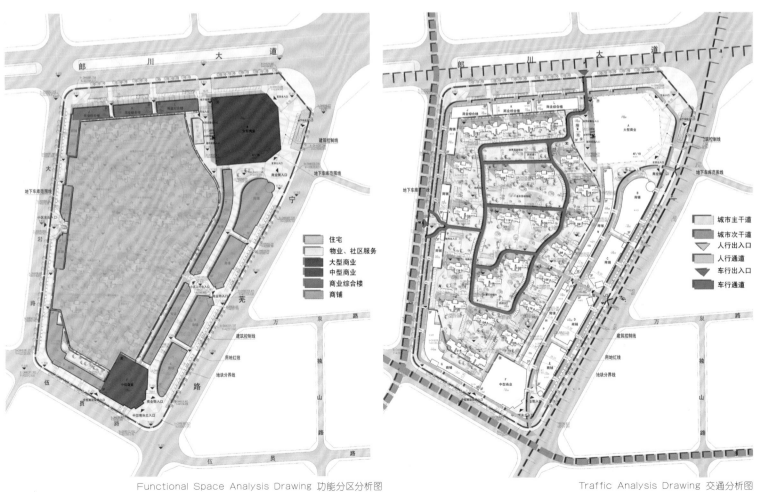

Functional Space Analysis Drawing 功能分区分析图

Traffic Analysis Drawing 交通分析图

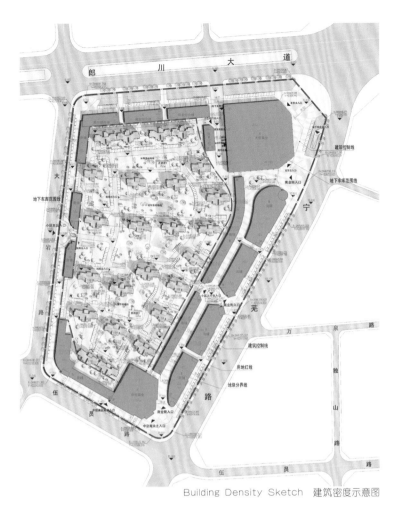

Building Density Sketch 建筑密度示意图

Building Green Coverage Rate Sketch 建筑绿地率示意图

Fire Analysis Diagram 消防分析图

Vertical Analysis Diagram 竖向分析图

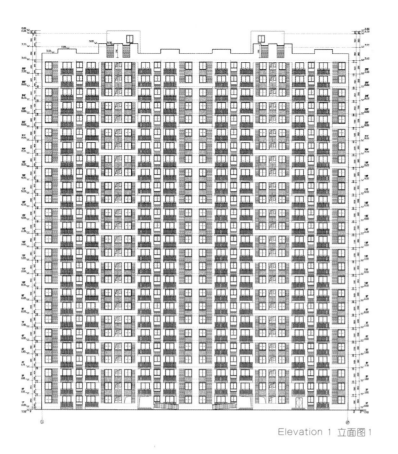

Elevation 1 立面图 1   Elevation 2 立面图 2

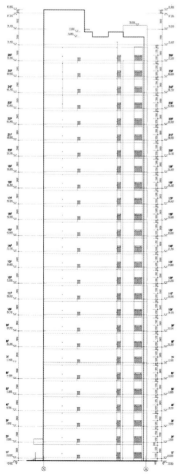

Elevation 3 立面图 3

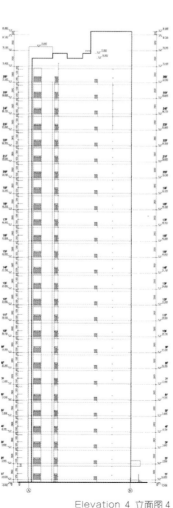

Elevation 4 立面图 4

■ Design Objective

Revolves around the people-oriented idea, the project aims to establish an ecological commercial and residential environment and create a section with cultural and local features where the layout is reasonable, function is complete, scene is green and life is convenient. It focuses on ecological environment, allocates resources appropriately and takes improving the quality of the environment as the basic starting point and ultimate goal, which fully embody the idea of sustainable development.

■ 设计目标

设计围绕以人为本的思想，以建设生态型商业和居住的环境为目标，创造一个布局合理、功能齐备、绿意盎然、生活方便以及具有文化内涵和当地特色的区域。注重生态环境，合理分配各项资源，全面体现可持续发展的思想，把提高环境质量作为规划设计的基本出发点和最终目的。

■ General Layout

In terms of general layout, the clear function layout not only meets the commercial and residential requirements, but also is in line with the specifications of urban planning, fire protection, sunlight and ventilation.

■ 总平面布局

在总平面布局的设计方面，功能分区明确，满足商业、居住的功能要求，同时也符合城市规划、消防、日照及通风等规范要求。

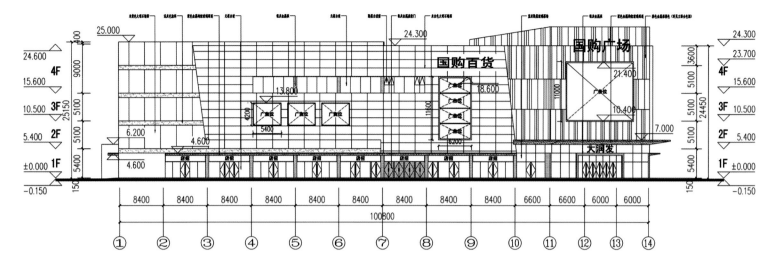

Elevation 5 立面图 5

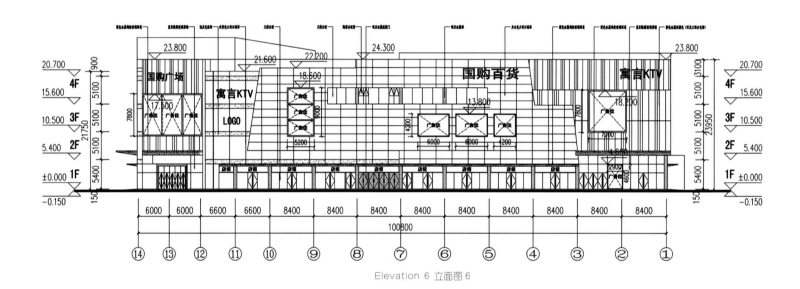

Elevation 6 立面图 6

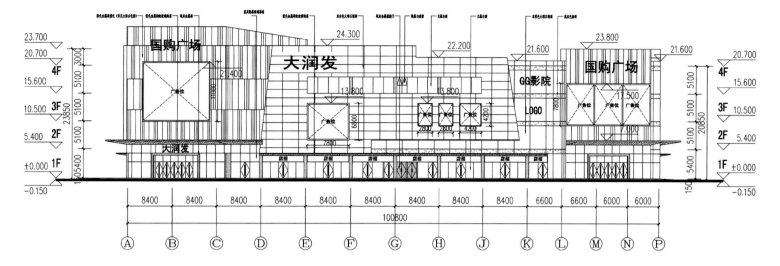

Elevation 7 立面图 7

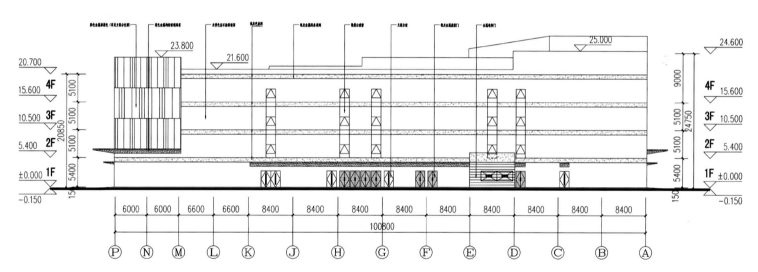

Elevation 8 立面图 8

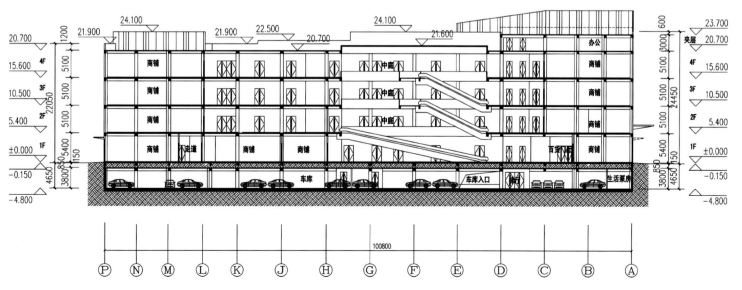

Section 1-1 1-1 剖面图

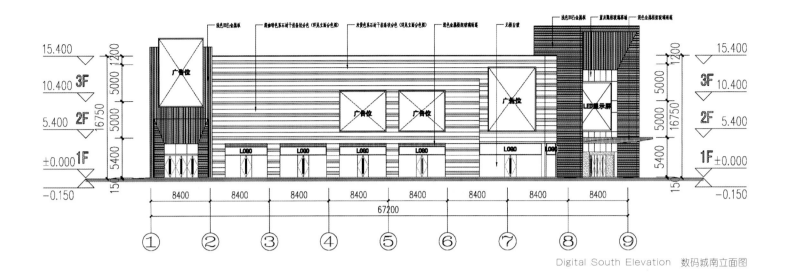

Digital South Elevation 数码城南立面图

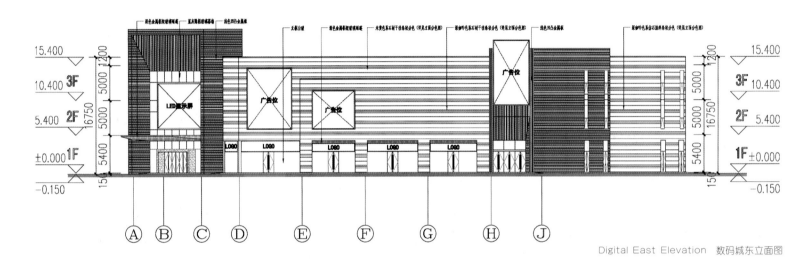

Digital East Elevation 数码城东立面图

Side Elevation 1 轴立面图1

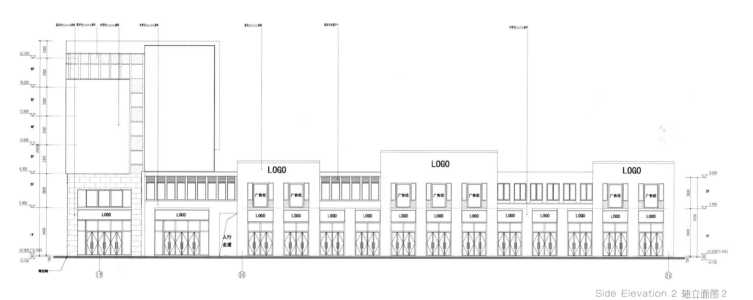

Side Elevation 2 轴立面图2

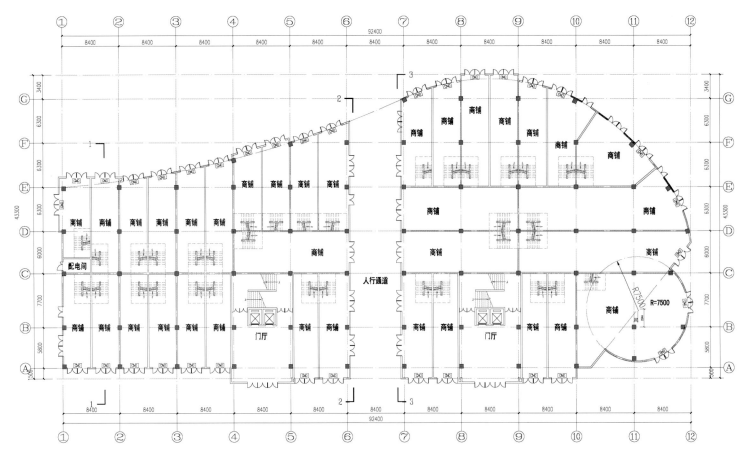

Second Floor Plan of Commercial 商业一层平面图

■ Green Design

As a major feature of the project's design, the "green community" is planned to be efficient, comfort, energy-saving, environment-friendly and ecological balanced. By creating a close-to-green space, residents could make the most of the green area.

■ 绿色设计

营造"绿色社区"是本设计的一大特点，以求达到高效舒适、节能环保、生态平衡的目的。亲绿空间的创造，最大限度地使人进入绿地。

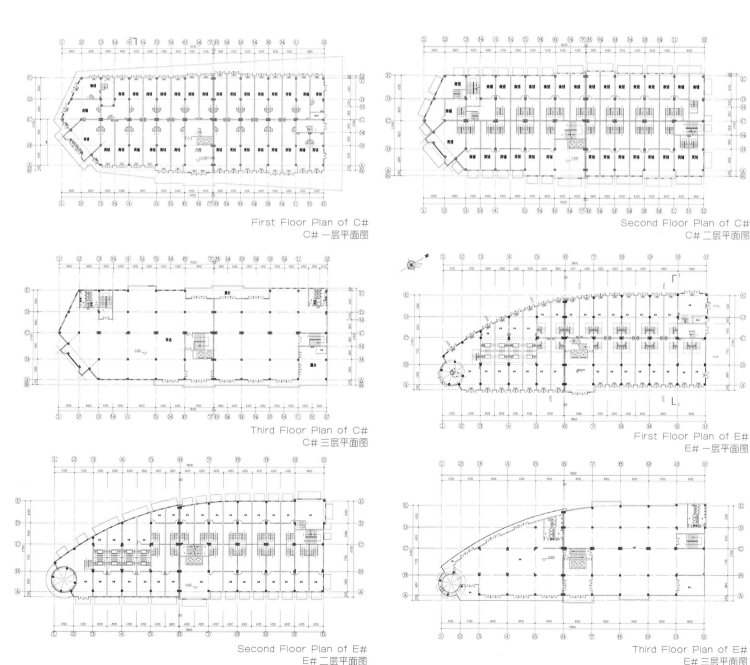

First Floor Plan of C#
C# 一层平面图

Second Floor Plan of C#
C# 二层平面图

Third Floor Plan of C#
C# 三层平面图

First Floor Plan of E#
E# 一层平面图

Second Floor Plan of E#
E# 二层平面图

Third Floor Plan of E#
E# 三层平面图

Summer International Retail and Entertainment Center

世邦国际商贸中心

Keywords 关键词
Building Blocks 积木块
Stone Wall 石级墙
Cantilever Structure 悬挑结构

Features 项目亮点

The design takes inspiration from its location and topography with a dynamic "urban super wall" defining the sites edge with the urban grid.

项目构筑了一座极具动感的"超级城墙",以仿"积木块"的形式,沿着围墙区域散落,分隔项目的地界与城市的框架。

Location: Zhuhai, Guangdong, China
Client: SUMMER INDUSTRIAL GROUP
Architectural Design: 10 DESIGN
Architects: Gordon Affleck
Architectural Design Team: Jamie Webb, Bernice Kwok, Kevis Wong, Nick Chan, Jason Easter
Landscape Design Team: Christian Dierckxsens, Ewa Koter, Alicia Johannesen
Land Area: 170,000 m^2
Floor Area: 510,000 m^2

■ **Overview**

The mixed use development in Zhuhai, China contains 360,000 m^2 leasable retail space together with commercial, hotel serviced apartment and residential accommodation totaling 510,000 m^2 of accommodation.

■ **项目概况**

坐落于中国珠海的世邦国际商贸中心(暂定名)是一个综合用途项目,包括360 000 m^{2}的零售空间,加上项目的商用部分、酒店式服务公寓及住宅区,共构成510 000 m^{2}的总建筑面积。

项目地点:中国广东省珠海市
业　主:世邦集团
设计单位:拾稼设计
设　计　师:艾高登
建筑设计团队:Jamie Webb, Bernice Kwok, Kevis Wong, Nick Chan, Jason Easter
景观设计团队:Christian Dierckxsens, Ewa Koter, Alicia Johannesen
占地面积:170 000 m^2
建筑面积:510 000 m^2

■ **Architecture Design**

The site of the development is unique in this important growing city, as it is the meeting point between the grid of the city and the natural topography of the surrounding hill range. The design of the development takes inspiration from this with a dynamic "urban super wall" defining the sites edge with the urban grid. The super wall is made up of series of giant, stacked stone, steel and LED blocks that are stacked to open and cantilever out across the street.

The LED blocks acts as media and light entry gateways and break out points. These breaks gates within the wall reveal the softer organic, planted terrace building and street forms within like a giant secret garden. The terraces within the heart of the site are sculpted to reflect the flow of pedestrian movement through the site along undulating terraced valleys that open to create external plazas and close to create intimate shaded courtyards.

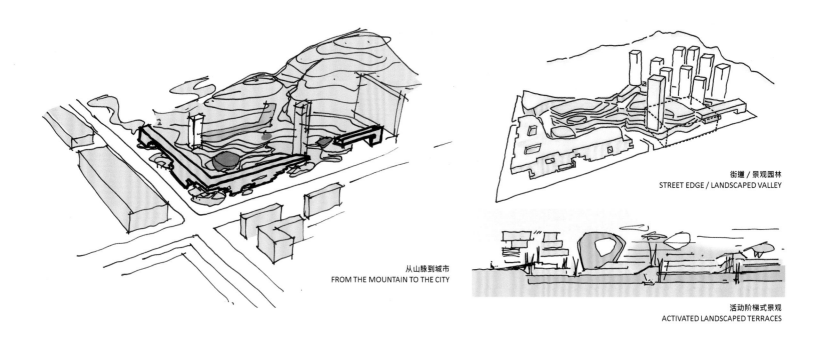

从山脉到城市
FROM THE MOUNTAIN TO THE CITY

街道／景观园林
STREET EDGE / LANDSCAPED VALLEY

活动阶梯式景观
ACTIVATED LANDSCAPED TERRACES

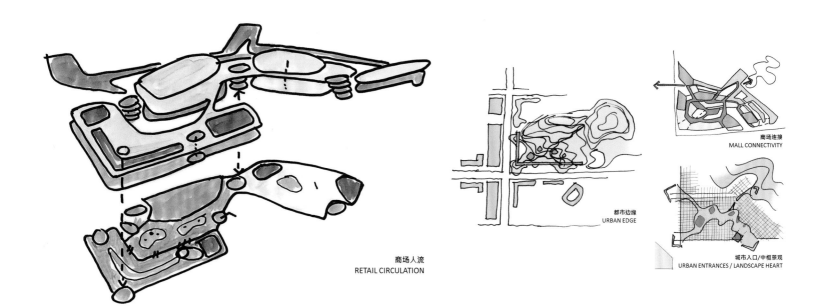

商场人流
RETAIL CIRCULATION

都市边缘
URBAN EDGE

商场连接
MALL CONNECTIVITY

城市入口／中枢景观
URBAN ENTRANCES / LANDSCAPE HEART

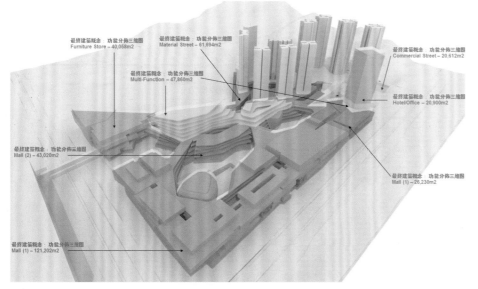

■ 建筑设计

本项目的位置拥有不可多得的优越性，它既位于城市主干道汇聚之处，同时也与周边的自然山脉互相交汇。项目以此为设计灵感，构筑一座极具动感的"超级城墙"，以仿"积木块"的形式，沿着围墙区域散落，分隔项目的地界与城市的框架。其中，带有不同设计形式的积木墙穿插在自然的景观花园之中，通过巨型的石级墙，或不锈钢及LED发光积木块，塑造了穿越街道的开敞及悬挑结构。

作为建筑门户的通道与泛光照明，LED发光积木块也是"超级城墙"的分隔点。这些巨墙之间的空隙，透露出内部植栽柔软有机的绿化台阶建筑和街道，形成一座巨型的神秘花园。而在项目中心区的台阶建筑，以雕塑造型反映行人活动流线。沿起伏的台阶建筑山谷空间，或打开形成开放式广场，或收缩形成私密的阴凉庭院。

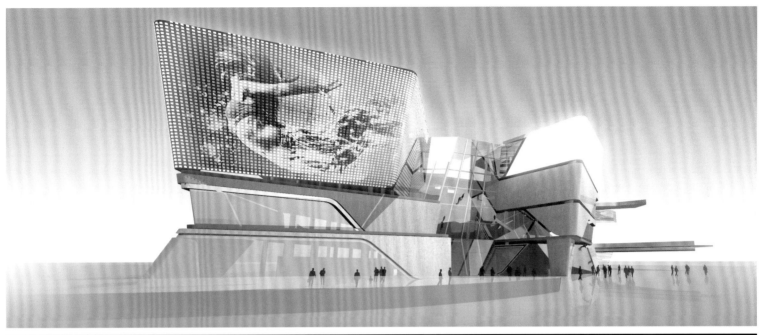

Qingdao Urban Balcony
青岛城市阳台

Keywords 关键词

Curve Block
曲线形体

Separated Basement
底层分离

Connected Roof
顶部相连

Features 项目亮点

Inspired by the mountains and conches, the architectures build a curve block with horizontal lines just like the veins of the mountains. Meanwhile, in order to avoid to create a bloated building, the architectures separate the middle of the building and then connect the roof.

受山峰和海螺的启发，将整个建筑体形设计为曲线，层层的横线条如同山峰的叠峦。同时为避免建筑臃肿庞大，把建筑中间分开，只在顶部相连，看似分离，实为一体。

Location: Qingdao, Shandong, China
Developer: Shimao Group
Architectural Design: Dada (Beijing) Architetural And Planning CO. Limitied
Site Area: 86,940 m²
Total Floor Area: 187,879 m²

■ **Overview**

This project has the advanced location, backing by modern metropolis, and facing the ocean. A sense of busy, fast-paced and crowned provided by the modern metropolis, while the wide ocean is able to calm them down. A landmark building located here need to express the characters of both, which is the combination of contradiction and identity.

项目地点：中国山东省青岛市
开 发 商：世茂集团
设计单位：DADA大的建筑设计咨询（北京）有限公司
基地面积：86 940 m²
总建筑面积：187 879 m²

■ **项目概况**

本项目占有独特的地理优势：背靠现代都市，面向孕育生命的海洋。现代都市给人以繁忙、快节奏拥挤的感觉，而广阔海洋能给人以沉静。地标建筑要同时彰显两者，本身就是矛盾与统一的结合体。

■ Design of the Block

The stacked mountains are the source of dynamic which can pass on the power to people. While the conches come from ocean, with graceful curve but tenacious body, is the perfect combination of morbidezza and tenacity. Inspired by these two elements, the architectures build a curve block with horizontal lines just like the veins of the mountains. Meanwhile, because of the huge size, in order to avoid to create a bloated building, the architectures separate the middle of the building and then connect its roof as the conches, and a new channel for the ocean view is created at the same time.

■ 建筑形体设计

层叠的山峰给人以力量,是动力的源泉。海螺生于海洋,其身体不但曲线柔美而且又坚韧不屈,是柔美与刚韧完美结合的综合体。受两者的启发将整个建筑体形设计为曲线,层层的横线条如同山峰的叠峦。因建筑身高体长,为避免臃肿庞大,受海螺体形的启发,在建筑中间分开,只在顶部相连,看似分离,实为一体,同时巧妙地为都市开拓一海景视线通道。

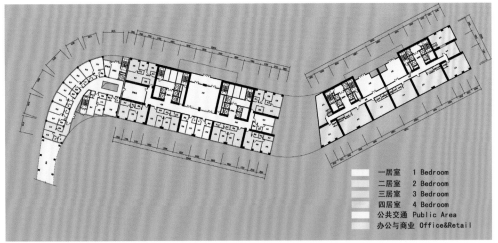
Main Building First Floor Plan 主楼首层平面图

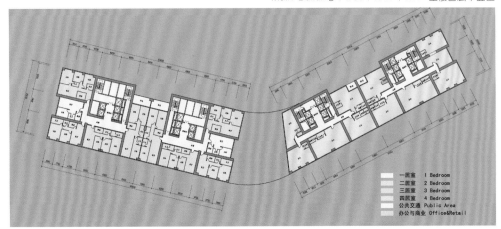
Main Building Standard Layer Plan 主楼标准层平面图

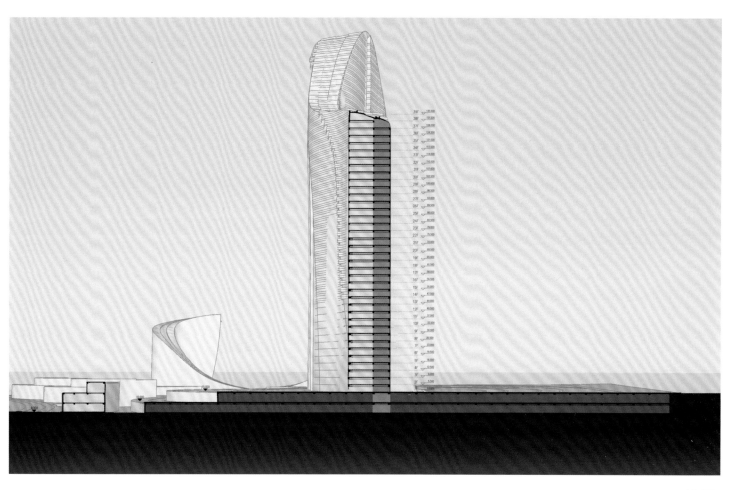

Section 1 剖面图1

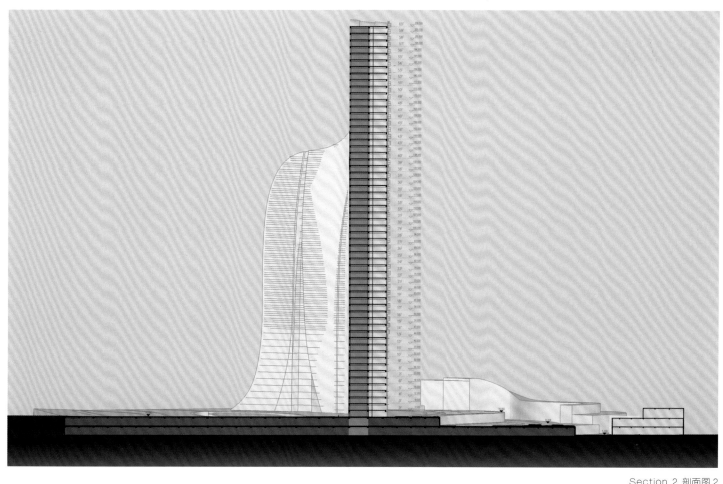

Section 2 剖面图2